WITHDRAWN

The Psychobiology of Anorexia Nervosa

Edited by
K. M. Pirke and D. Ploog

With 57 Figures

Springer-Verlag
Berlin Heidelberg New York Tokyo 1984

Privatdozent Dr. KARL MARTIN PIRKE
Professor Dr. DETLEV PLOOG

Max-Planck-Institut für Psychiatrie
Kraepelinstraße 10, D-8000 München 40

ISBN 3-540-13196-5 Springer-Verlag Berlin Heidelberg New York Tokyo
ISBN 0-387-13196-5 Springer-Verlag New York Heidelberg Berlin Tokyo

Library of Congress Cataloging in Publication Data. The psychobiology of anorexia nervosa. Bibliography: p. 1. Anorexia nervosa − Physiological aspects. 2. Psychobiology. I. Pirke, K. M. (Karl Martin), 1941− . II. Ploog, Detlev, 1920− . [DNLM: 1. Anorexia nervosa − Congresses. WM 175 P974 1983] RC552.A5P79 1984 616.85'2 84-1358 ISBN 0-387-13196-5 (U.S.)

This work is subject to copyright. All rights are reserved, whether the whole or part of the material is concerned, specifically those of translation, reprinting, re-use of illustrations, broad-casting, reproduction by photocopying machine or similar means, and storage in data banks. Under § 54 of the German Copyright Law where copies are made for other than private use, a fee is payable to "Verwertungsgesellschaft Wort", Munich.

© by Springer-Verlag Berlin Heidelberg 1984
Printed in Germany

The use of registered names, trademarks, etc. in this publication does not imply, even in the absence of a specific statement, that such names are exempt from the relevant protective laws and regulations and therefore free for general use.

Product Liability: The publisher can give no guarantee for information about drug dosage and application thereof contained in this book. In every individual case the respective user must check its accuracy by consulting other pharmaceutical literature.

Typesetting, printing and bookbinding: v. Starck'sche Druckereigesellschaft m.b.H., Wiesbaden
2121/3130-543210

Contents

The Importance of Physiologic, Metabolic, and Endocrine Studies for the Understanding of Anorexia Nervosa
D. Ploog. With 2 Figures 1

Nutritional Control of Central Neurotransmitters
R. J. Wurtman and J. J. Wurtman. With 2 Figures 4

Impaired Control of Appetite for Carbohydrates in Some Patients with Eating Disorders: Treatment with Pharmacologic Agents
J. J. Wurtman and R. J. Wurtman. With 1 Figure 12

Animal Models: Anorexia Yes, Nervosa No
N. Mrosovsky. With 1 Figure 22

Noradrenergic Function in the Medial Hypothalamus: Potential Relation to Anorexia Nervosa and Bulimia
S. F. Leibowitz 35

Effect of Starvation on Central Neurotransmitter Systems and on Endocrine Regulation
K. M. Pirke, B. Spyra, M. Warnhoff, I. Küderling, G. Dorsch, and C. Gramsch. With 9 Figures 46

Neurotransmitter Metabolism in Anorexia Nervosa
M. H. Ebert, W. K. Kaye, and P. W. Gold. With 5 Figures 58

Sleeping and Waking EEG in Anorexia Nervosa
D. J. Kupfer and C. M. Bulik 73

Gastric Function in Primary Anorexia Nervosa
A. Dubois, H. A. Gross, and M. H. Ebert. With 3 Figures 87

Psychophysiological Indices of the Feeding Response in Anorexia Nervosa Patients
R. Hölzl and S. Lautenbacher. With 13 Figures 93

Endocrine Function in Magersucht Disorders
P. J. V. Beumont. With 1 Figure 114

Hypothalamic Pituitary Function in Starving Healthy Subjects
M. M. Fichter and K. M. Pirke. With 5 Figures 124

Perceptions of the Body in Anorexia Nervosa
P. E. Garfinkel and D. M. Garner. With 1 Figure 136

Treatment of Anorexia Nervosa: What Can Be the Role of Psychopharmacological Agents?
A. H. Crisp. With 8 Figures 148

The Basis of Naloxone Treatment in Anorexia Nervosa and the Metabolic Responses to It
I. H. Mills and L. Medlicott. With 3 Figures 161

PET Investigation in Anorexia Nervosa: Normal Glucose Metabolism During Pseudiatrophy of the Brain
H. M. Emrich, J. J. Pahl, K. Herholz, G. Pawlik, K. M. Pirke, M. Gerlinghoff, W. Wienhard, and W. D. Heiss. With 2 Figures .. 172

Concluding Remarks
K. M. Pirke and P. Ploog 179

List of Contributors

You will find the addresses at the beginning of the respective contribution

Beumont, P. J. V.   114
Bulik, C. M.   73
Crisp, A. H.   148
Dorsch, G.   46
Dubois, A.   87
Ebert, M. H.   58, 87
Emrich, H. M.   172
Fichter, M. M.   124
Garfinkel, P. E.   136
Garner, D. M.   136
Gerlinghoff, M.   172
Gold, P. W.   58
Gramsch, C.   46
Gross, H. A.   87
Heiss, W. D.   172
Herholz, K.   172
Hölzl, R.   93

Kaye, W. K.   58
Küderling, I.   46
Kupfer, D. J.   73
Lautenbacher, S.   93
Leibowitz, S. F.   35
Medlicott, L.   161
Mills, I. H.   161
Mrosovsky, N.   22
Pahl, J. J.   172
Pawlik, G.   172
Pirke, K. M.   46, 124, 172, 179
Ploog, D.   1, 179
Spyra, B.   46
Warnhoff, M.   46
Wienhard, W.   172
Wurtman, J. J.   4, 12
Wurtman, R. J.   4, 12

Acknowledgements

We are grateful for the support of the Deutsche Forschungsgemeinschaft (German Research Foundation) and the Max-Planck-Gesellschaft, which allowed us to organize the workshop on Psychobiology of Anorexia Nervosa at Ringberg Castle on 7 and 8 July 1983. Additional financial support was provided by Merck AG, Darmstadt, Federal Republic of Germany.

Mrs. Betty Weyerer's valuable assistance allowed us to prepare this volume for publication in a rather short amount of time. Drs. J. Pahl, M. Warnhoff, I. Küderling, E. Philipp, and U. Schweiger assisted in the organization of the meeting at Ringberg Castle. The excellent cooperation with Springer-Verlag is gratefully acknowledged.

Munich, August 1983   K. M. Pirke   D. Ploog

The Importance of Physiologic, Metabolic, and Endocrine Studies for the Understanding of Anorexia Nervosa

D. Ploog[1]

Anorexia nervosa is a fascinating and most complicated disease insofar as mental, as well as somatic, functions are considerably changed and impaired by it. In this sense it is a truly psychosomatic disease. It also possesses all the characteristics of an addiction as far as course and outcome, coping strategies, subjective feelings, and psychopathology are concerned (Ploog 1981). A deviant psychic development, usually starting around puberty, becomes evident in a growing rejection of food. This in turn leads to a reduction in weight and a state of malnutrition marked by a number of serious somatic symptoms. Several of these symptoms, such as the drop in temperature, the discontinuance of reproductive functions, and changes in the intermediate metabolism, are indicative of the existence of that physical state in which the remaining energy potential is used as sparingly as possible in order to prolong survival. These observations are based on a large number of case studies in which physiologic, metabolic, and endocrine malfunctions in anorexia nervosa patients were investigated.

It was the aim of this symposium on the psychobiology of anorexia nervosa to search for the answers to three important questions. The first question is whether all somatic symptoms in anorexia nervosa patients are caused by the malnutrition or whether there are, in addition, endocrine and other malfunctions which must be excepted to be direct causes of anorexia nervosa. It may be assumed that if the state of malnutrition has such a marked effect on the form and function of the body it will also affect the central nervous system (CNS). We know, of course, that the wasting of cell tissue due to starvation does not affect the brain for quite a while. For example, the protein content of the CNS is not reduced when the body is starving, but the muscular system and all of internal organs lose considerable quantities of protein. Nevertheless, changes in the CNS have been registered by several authors. As an example, Fig. 1 shows a considerable shrinkage of brain contour. Aside from morphologic changes, however, we must also expect to find biochemical changes in the brain, especially since we know that the activity of specific central neurotransmitter systems depends on the availability of certain amino acids in the blood (see R. J. Wurtman and J. J. Wurtman, this volume).

The second question is, how do malnutrition and anorexia nervosa change the cerebral metabolism? We know from many studies that malnutrition alone can change cerebral functions, such as behavior, disposition, and motivation. Hilde Bruch has already pointed out that the bizarre treatment of food that is typical for anorexia nervosa patients has also been observed in patients suffering

---

1 Max-Planck-Institut für Psychiatrie, Kraepelinstraße 10, D-8000 München 40

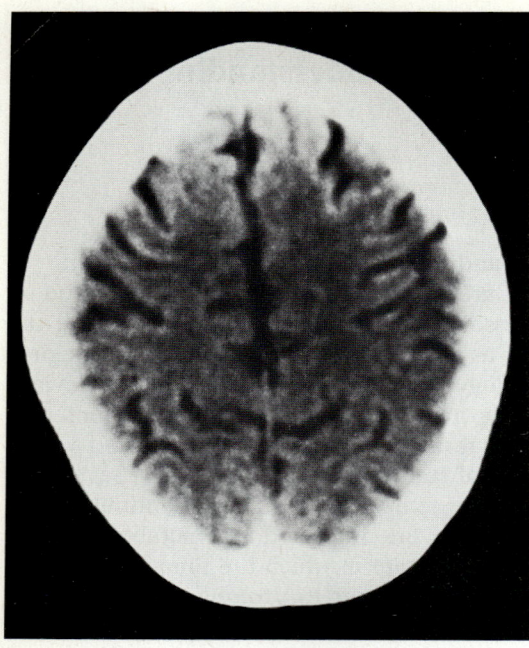

Fig. 1. CT scan displays a moderate widening of cortical sulci and of the frontal interhemispheric fissure

from malnutrition. Fasting experiments conducted in Munich (Fichter et al., this volume), have shown that even a normally healthy person, when fasting, will experience the stages of depression typical for many anorexia nervosa patients. We are still far from being able to detect a clear causative connection between psychic changes and biochemical malfunctions in the CNS. There are two possible means of answering this question as to how the cerebral metabolism is changed by malnutrition and anorexia nervosa. The first is biochemical examinations of the brain of a malnourished animal. The second possibility is based on our knowledge, however fragmentary, of the role specific neurotransmitter systems play in the regulation of sleep, for instance, or in temperature or neuroendocrine functions. Several of the contributions to this volume describe changes in the starving brain. These reports will raise questions as to whether we are already entitled to form hypotheses regarding biochemical and functional changes in the CNS on the basis of the phenomena outlined.

The third, and perhaps most important, question is whether knowledge about the effects of malnutrition on the body and psyche of an anorectic patient simply improves our understanding of the symptoms of the disease, or whether, in addition, it allows insights into the phenomena essential for understanding the pathogenesis of the disease. Figure 2 shows a hypothetical description of the development of anorexia nervosa. The hypothesis is based on the assumption that the effect of malnutrition on a number of peripheral and central physical functions keeps the disease alive, and even aggravates it in the manner of a vicious circle. The changes observed in the stomach motility of a malnourished

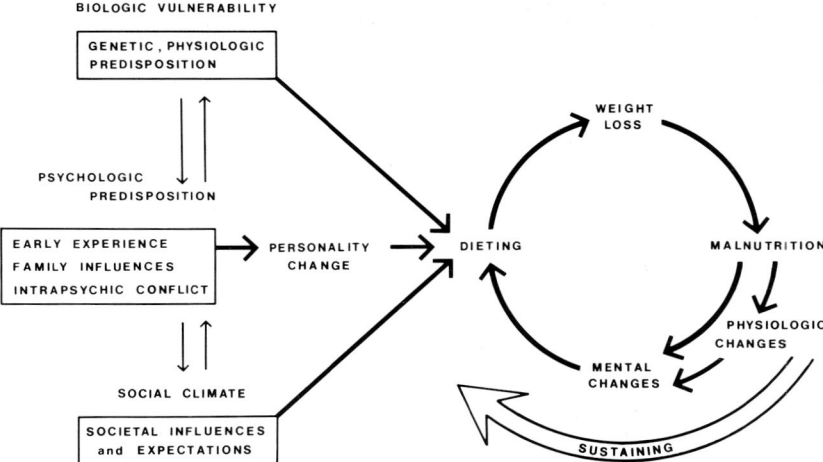

Fig. 2. Hypothetical causes and the "vicious circle" of anorexia nervosa. For explanation see text. (Adapted from Lucas 1981)

person (see Dubois et al. and Hoelzl and Lautenbacher, this volume) are a good example of such mechanisms. They cause an exaggerated feeling of fullness in the patient, who consequently reduces his food intake. Another example is social withdrawal. It has been observed in anorexia and fasting cases and can eventually result in the patient's complete surrender to the disease. An evaluation of the influences of psychic and physiologic changes induced by malnutrition on the course of anorexia nervosa was thus a matter of particular concern at this symposium.

On the basis of these considerations, a therapeutic aspect was dealt with: Can pharmacological means be applied to correct peripheral and central malfunctions for the purpose of interrupting the vicious circle, preparing the patient for psychotherapy, and perhaps changing his eating habits and his appetite? We have not found the answers to all the questions asked here. By sharing our experiences, our knowledge, and our expectations, we have learned more about anorexia nervosa and return to our work inspired by new ideas and new hopes.

References

Lucas AR (1981) Towards the understanding of anorexia nervosa as a disease entity. Mayo Clin Proc 56: 254–264

Ploog D (1981) Neuroethological aspects of anorexia nervosa. In: Perris C, Struwe G, Jansson B (eds) Biological psychiatry. Elsevier, Amsterdam, pp 1027–1034

Nutritional Control of Central Neurotransmitters

R. J. Wurtman and J. J. Wurtman[1]

Some of the changes in plasma composition which follow eating (e.g., those in the concentrations of the neutral amino acids and of choline) can have important secondary effects on the nervous system, modulating both the rates at which particular neurons convert nutrients to neurotransmitters, and the quantities of neurotransmitters released when the neurons fire (Wurtman 1982; Wurtman et al. 1980). Moreover, these effects can be amplified by administering the nutrients in pure form, as though they were drugs (Wurtman 1982; Wurtman et al. 1980) and, in the case of amino acids, by mixing them with carbohydrates (Mauron and Wurtman 1982). The brain apparently takes advantage of its ability to couple plasma composition with neurotransmission to obtain information

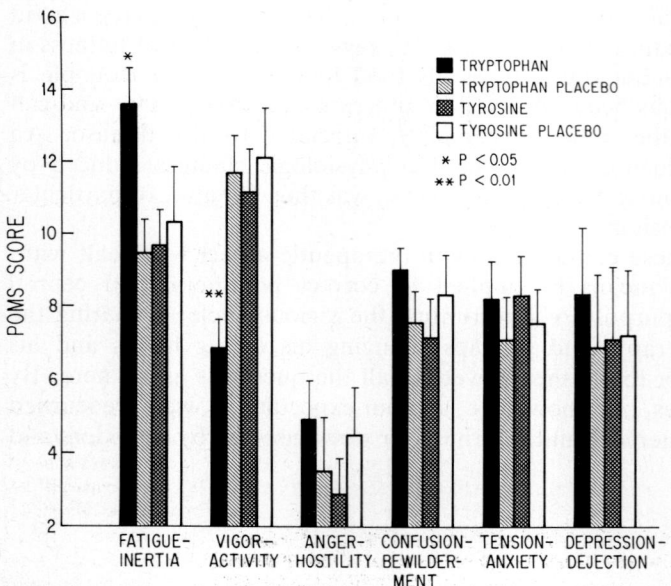

Fig. 1. Effects of tryptophan or tyrosine on mood. Subjects received oral tryptophan (50 mg/kg) or tyrosine (100 mg/kg), using a double-blind, placebo-controlled, crossover design. The substances were ingested at 07:00 h and behavioral testing began at 09:00 h. Mood, measured by the Profile of Mood States (POMS), was significantly altered. Specifically, tryptophan significantly decreased Vigor ($P < 0.01$) and increased Fatigue ($P < 0.05$) compared with either placebo or tyrosine. (H. R. Leiberman, personal communication)

---

[1] Massachusetts Institute of Technology, Department of Nutrition and Food Science, Cambridge, MA 02139, USA

about the individual's metabolic state and to make decisions about subsequent food intake and about cyclic behavioral processes such as sleeping (Wurtman RJ 1983). For example, if a meal is rich in carbohydrate and poor in protein, postprandial changes in plasma amino acids concentrations cause brain tryptophan levels to rise; this, in turn, accelerates the production and release of tryptophan's neurotransmitter product serotonin (Fernstrom and Wurtman 1971; Fernstrom and Wurtman 1972), which predisposes the individual to choose away from carbohydrates and towards protein at the next meal. It may also make him a little sleepy, and a little more likely to make errors in certain types of work-related tasks (Spring et al. 1983; Fig. 1). In contrast, a protein-rich meal,

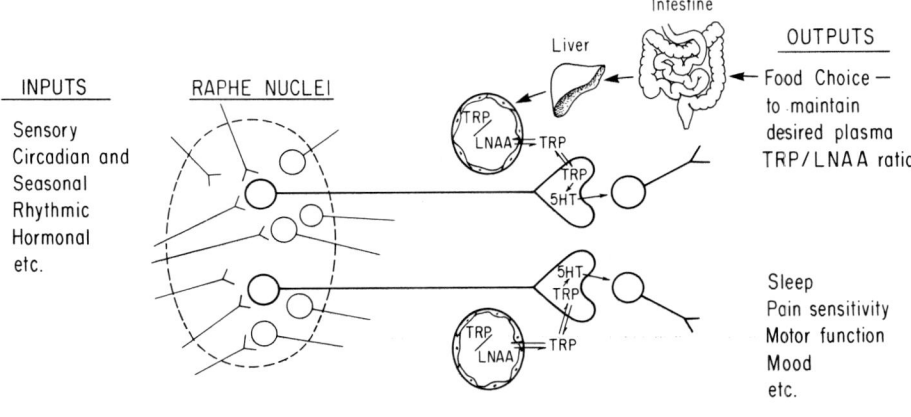

TOTAL OUTPUT OF SEROTONIN (5-HT) VARIES WITH:

1. Number of 5-HT nerve terminals (? Genetic);
2. Average firing frequency of 5-HT neurons;
3. Average number of 5-HT molecules released per terminal per firing (partly nutritional).

Fig. 2. Effects of plasma composition and presynaptic inputs on serotonin (5HT) release. Serotonin-releasing neurons, with cell bodies in the raphe nuclei, receive presynaptic inputs from elsewhere in the brain, providing them with signals relating to sensory inputs, circadian and seasonal rhythms, and hormonal effects. Their axons distribute widely in the brain, releasing serotonin into synapses and, probably, in the brain's extracellular space. Within serotoninergic nerve terminals, synthesis of the transmitter is affected by available tryptophan (TRP) levels; these, in turn, depend on the concentrations of TRP and of other competing LNAA in the brain's capillaries. Consumption of a carbohydrate-rich, protein-poor meal increases the "plasma/TRP ratio," thus enhancing serotonin synthesis. One consequence of the resulting increase in serotoninergic neurotransmission is to decrease the individual's subsequent consumption of carbohydrate relative to protein, which has the effect of restoring the plasma/TRP ratio and brain serotonin release to "desired" levels. Another potential consequence is to enhance *any* physiological or behavioral mechanism involving serotonin, e.g., sleep onset. The total amount of serotonin released in the brain per unit of time depends both on the number of serotoninergic terminals and on the firing frequencies of the neurons and the amount of transmitter released per terminal. Brains containing reduced numbers of serotoninergic neurons might be expected to be less responsive to food-induced changes in the plasma/TRP ratio

which diminishes brain serotonin, can have opposite effects (Fernstrom and Wurtman 1972). The biochemical mechanisms underlying the nutritional control of neurotransmitter synthesis and food intake are discussed below (c.f. Fig. 2).

The changes in brain serotonin induced by consuming foods of varying protein-to-carbohydrate ratios might also be expected to influence any behavioral phenomena mediated in part by serotonin-releasing neurons, for example mood. Moreover, the consequences of these changes might be expected to be especially pronounced in people with brain disorders involving serotonin-releasing neurons, such as depression. Other neurotransmitters besides serotonin are also affected by nutrient availability, but behavioral effects have not been so clearly shown to occur in normal people following postprandial changes in their synthesis. These other neurotransmitters (principally the catecholamines, formed from tyrosine, and acetylcholine, formed from choline) are affected by nutritional state only under particular circumstances, that is, only when the particular catecholaminergic or cholinergic neuron in which they are formed happens to be firing relatively frequently. (As a corollary, the brain can cause these neurons to become *un*responsive to administration of the precursor, just by slowing their firing frequency.) In contrast, the release of serotonin always appears to be affected by treatments that raise or lower brain tryptophan. The abilities of exogenous tyrosine or choline to act as amplifiers, increasing the amounts of their neurotransmitter products being released from physiologically active neurons when the brain chooses to allow them to do so, affords these compounds considerable potential use in treating diseases related to inadequate catecholamine or acetylcholine release: Unlike drugs that directly stimulate postsynaptic receptors, they should be expected to produce few if any side-effects related to inappropriate increases in neurotransmitter output.

Food Consumption and Plasma Amino Acid Pattern

Foods are, ultimately, mixtures of chemicals, which are consumed electively at intervals during the day, partially metabolized within the gut, and then released into the bloodstream and taken up into the tissues. Some of these chemicals, the essential nutrients, are continuously required by cells; at times when foods are not being digested and absorbed these compounds are released into the bloodstream from reservoirs in particular tissues (e.g., proteins for tryptophan; membrane phospholipids for choline). The rates at which major nutrients pass from the bloodstream into most tissues are controlled by hormones released postprandially, especially insulin, which facilitates the passages of glucose, fatty acids, and most amino acids. (The fall in plasma insulin that occurs when foods stop being absorbed causes the net flow of these nutrients to reverse, and their plasma levels to rise.) Brain is an exception: the passage of nutrients between *its* extracellular space and the plasma appears not to be directly affected by such hormones as insulin, depending rather on plasma nutrient levels and on the kinetic characteristics of specific transport systems located in the endothelial cells lining the brain's capillaries. The macromolecules which comprise these systems physically carry the nutrient molecules in either direction across the blood-brain barrier (Pardridge 1977). These macromolecules that transport the neurotransmitter precursors normally are unsaturated with their circulating ligands, and hence a postprandial increase (or decrease) in the plasma concentration of a nutrient like tyrosine or choline will facilitate (or slow) the compound's uptake into the brain.

Postprandial changes in the plasma levels of most nutrients are relatively short lived, since the compounds can be metabolized by the gut and liver before entering the systemic circulation (and during recirculations), or incorporated within the tissues into larger, water-soluble, "reservoir" molecules such as glycogen, membrane phosphatides, triglycerides, and proteins. Plasma levels of some nutrients (e.g., calcium or glucose) are kept within narrow ranges postprandially, regardless of the composition of the food that has just been consumed, by homeostatic mechanisms; typically a small change in the nutrient's plasma level is "sensed", and processes are then activated which accelerate or slow its removal from the plasma by metabolism or tissue uptake. However, for other nutrients (the amino acids and choline) (Fernstrom et al. 1979; Hirsch et al. 1978), it seems that no such feedback loops operate. Plasma levels of these nutrients can thus vary across a wide range, depending solely on the composition of the food currently being digested. (For example, plasma valine levels may be as much as six-fold higher after a protein-rich breakfast than after a protein-free one, and plasma choline is elevated three-fold by a breakfast and lunch of eggs.)

Brain Serotonin Responses to Dietary Carbohydrate and Protein

Consumption of a carbohydrate-rich, protein-free breakfast causes major changes in the pattern of amino acids in the plasma, largely because of the secretion of insulin; the pancreatic hormone facilitates the uptake of most of the amino acids into such tissues as skeletal muscle, thus lowering their plasma concentrations markedly. Plasma tryptophan levels, however, are not reduced – primarily because the bulk of the tryptophan (TRP) is loosely bound to circulating albumin, which retards its entry into peripheral tissues while allowing it to enter the brain (Madras et al. 1974). Hence the ratio of plasma TRP to the plasma concentrations of other large, neutral amino acids (LNAA) such as leucine, isoleucine, valine, tyrosine, and phenylalanine rises markedly. This ratio largely determines the concentration of TRP within the brain (Fernstrom and Wurtman 1972) because of the characteristics of the transport mechanism that carries TRP across the blood-brain barrier. This transport system is unsaturated with its amino acid ligand; hence an increase or decrease in plasma TRP levels will, all other things being equal, rapidly alter TRP's flux. Moreover, a single transport mechanism carries all of the LNAA – including TRP – competitively; thus an increase in the plasma TRP/LNAA ratio (caused postprandially, for example, by an insulin-mediated fall in the other LNAA) rapidly elevates TRP levels throughout the brain (Fernstrom and Wurtman 1971; Fernstrom and Wurtman 1972; Pardridge 1977).

Within those relatively few brain neurons that convert tryptophan to serotonin and use the indoleamine as their neurotransmitter, the rise in TRP quickly increases the substrate saturation of the serotonin-forming enzyme TRP hydroxylase, thereby increasing serotonin's synthesis, its absolute levels within nerve terminals, and its release into synapses and the brain's extracellular space each time the neurons fire (Fernstrom and Wurtman 1971; Fernstrom and Wurtman 1972).

If the initial meal is, instead, rich in protein, plasma TRP levels rise (Fernstrom et al. 1979) because some of the TRP molecules in the protein are able to traverse the liver and enter the bloodstream: levels of the other LNAA not only do not fall, however, as after a carbohydrate meal, but actually rise manyfold. The difference between the responses of plasma TRP and other plasma LNAA to dietary protein reflects their relative abundance in the protein. Tryptophan is scarce, comprising only about 1%−1.5% of most proteins; the other LNAA, as a group, are not. It also reflects the fact that some of the other LNAA in dietary proteins − leucine, isoleucine, and valine − are, unlike TRP, largely unmetabolized during their passage through the liver. Hence the plasma TRP/LNAA ratio falls in response to protein ingestion, as do brain TRP and serotonin levels (Fernstrom and Wurtman 1972). Serotonin-releasing neurons can thus be conceived of as "variable-ratio sensors" (Wurtman 1982), emitting more or less of their signal, serotonin, depending upon the chemicals being sensed (plasma concentrations of TRP and the other LNAA). These chemicals, in turn, vary predictably, depending upon the composition of the food currently being digested and absorbed.

Serotonin Neurons and the Control of Food Choice

If animals are allowed to choose concurrently among two or more foods of differing composition, their observed behavior suggests that appetitive mechanisms are operating which allow them to regulate not only the mass of food that they eat and its number of calories, but also the proportion of protein to carbohydrate (or carbohydrate to protein) within the meal (Wurtman et al. 1983; Theall et al., unpublished observations). This can be demonstrated either by keeping the carbohydrate (and calorie) contents of the test foods constant and varying the percentage of protein, or by keeping the protein (and calories) constant and varying the percentage of carbohydrate. (That is, one can demonstrate regulation of both the protein and the carbohydrate components of the diet.) The ability to choose among foods so as to obtain a "desired" proportion of carbohydrates is apparently independent of the sweetness of the carbohydrate being tested: regulation is as easy to demonstrate for dextrin as for dextrose or sucrose (Wurtman and Wurtman 1979).

Serotonin-releasing neurons are apparently key components of the brain mechanisms underlying this regulation of food choice: rats can be caused to choose away from carbohydrate (that is, to increase the protein/carbohydrate ratio of the test meal) either by giving them a carbohydrate-rich "premeal" (Wurtman et al. 1982) or by administering any of a large number of drugs which act at different loci to enhance serotonin-mediated neurotransmission (Wurtman and Wurtman 1979). Moreover, dietary manipulations (such as the chronic consumption of a carbohydrate-poor diet) which diminish the plasma TRP/LNAA ratio and thus reduce brain TRP and serotonin levels, sharply increase the proportion of carbohydrate that the animal subsequently eats when it is given the opportunity to choose (Wurtman et al. 1982). It seems not unlikely that the tendency of humans to consume a more or less constant proportion of their calories as protein, and the relative stability of lean body mass, may reflect the operation of this mechanism; it diminishes the likelihood that meals with imbalance in one direction (too much protein or carbohydrate) will be immediately followed by others with imbalance in the same direction.

It also seems possible that some of the common abnormalities in eating behavior — including those that can lead to obesity — may have their origin in disturbances in the mechanism through which the brain "learns" about the composition of the last meal (by nutrient-induced variations in serotoninergic neurotransmission) and "decides" what to eat next. Such disturbances could exist at several loci: for example, the plasma TRP/LNAA ratio might not rise after carbohydrate consumption because of inadequate insulin release (Heraief et al. 1983) or peripheral insulin resistance; the brain TRP/LNAA ratio might not respond appropriately to food-induced changes in the plasma TRP/LNAA ratio because of an abnormality in the blood-brain barrier transport system; the increase in brain tryptophan that follows a carbohydrate-rich, protein-poor meal might not have a sufficient effect on serotonin-mediated neurotransmission because the unfortunate individual has a brain containing too few serotoninergic neurons; the functional activity of the serotoninergic neurons may be disturbed by pathological processes arising elsewhere in the brain. While it is difficult to examine changes in brain chemistry using human subjects, it is relatively easy to determine whether, for example, obese individuals exhibit the same changes in the plasma TRP/LNAA ratio after a test meal as nonobese control subjects, and we are currently exploring such responses. During the past few years we have identified a subset of obese people whose problem seems to be specifically related to an inappropriate desire for carbohydrates, especially at certain times of day (Wurtman et al. 1981). If such "carbohydrate-cravers" (J. J. Wurtman 1983) compensate for their carbohydrate snacking by diminishing their mealtime food intake or by increasing their energy output they may avoid the development of obesity; however, if they eat normal-sized meals supplemented with multiple carbohydrate snacks they become obese. We find that administration of subanorectic doses of drugs (e.g., fenfluramine) which release brain serotonin can suppress the carbohydrate craving (Wurtman et al. 1981). It will be interesting to see whether supplemental TRP has a similar effect, especially in view of the evidence that the TRP/LNAA plasma ratio tends to be abnormally low in obese people and that the protein-rich, carbohydrate-poor diet (the "PSMF diet") with which many obese people are treated further lowers the this ratio (Heraief et al. 1983).

Other Nutrient-Dependent Neurotransmitters

As mentioned above, the rates at which physiologically active catecholaminergic neurons release their transmitter (dopamine, norepinephrine, or epinephrine) can be enhanced by giving the individual tyrosine, and can be suppressed by administering other LNAA (which compete with tyrosine for uptake into the brain). The effect of a given tyrosine dose (or of a given dose of TRP) can be potentiated by administering it concurrently with a carbohydrate source (Mauron and Wurtman 1982). The resulting secretion of insulin lowers plasma levels of the competing LNAA, and increases the proportion of administered molecules that are taken up into the brain (Mauron and Wurtman 1982; Pardridge 1977). The reason that tyrosine levels affect catecholamine synthesis only when neurons are firing rapidly has to do with the properties of the key enzyme, tyrosine hydroxylase, that converts the amino acid to the catecholamines (Wurtman 1982; Wurtman et al. 1980). When neurons fire frequently, the enzyme itself become phosphorylated; as a consequence its affinity for its cofactor, tetrahydrobioperin, increases markedly, so that it becomes limited by tyrosine. This phosphorylation is short-lived, so that soon after the neurons slow their firing they become unresponsive to additional tyrosine. Tyrosine administration can, theoretically, be useful in any experimental or clinical situation in which it would be desirable to have more catecholamine molecules released at a particular locus, inside or outside the brain. [Its uses in these situations, and the theoretical bases for such uses, have been reviewed extensively (Wurtman et al. 1980).]

Similarly, acetylcholine's production in and release from cholinergic neurons can depend on the availability of free choline, when particular cholinergic neurons happen to be firing frequently. Any treatment that increases plasma choline, including oral choline chloride, dietary phosphatidylcholine (PC; lecithin), or even eating eggs or liver, will have this effect. The primary reason that brain choline levels rise after choline consumption is not that more plasma choline enters the brain but that less leaves it. The brain is able to synthesize choline de novo (Blusztajn and Wurtman 1982); however, most of the choline is lost by secretion into the bloodstream via the operation of a bidirectional blood-brain barrier choline transport system. Raising plasma choline slows this loss, thereby allowing brain choline to accumulate. Choline is converted to acetylcholine by the enzyme choline acetyltransferase; like tyrosine hydroxylase, this enzyme is highly unsaturated with its substrates (choline and acetyl-CoA), and becomes highly responsive to treatments increasing brain choline levels when the neurons containing it fire frequently (Blusztajn and Wurtman 1983). The mechanism that couples the choline dependence of this enzyme to neuronal firing frequency awaits identification. Choline (or PC) administration has been found to be useful in treating tardive dyskinesia, mania, and several other neuropsychiatric diseases (Wurtman et al. 1980). Given chronically, PC may also constitute a useful adjunct in the management of Alzheimer's disease, or senility (Levy et al. 1983); given acutely, it may potentiate the therapeutic effects of such drugs as physostigmine or piracetam, which enhance central cholinergic transmission and which are being widely tested in this disorder.

Although the effects of tyrosine and choline on brain function are less ubiquitous than those of TRP, when they do occur (in rapidly firing neurons) they can be of major importance for health, for example, in memory, or in control of mood, or in sustaining normal cardiovascular function. Hence it seems prudent that care should be given to satisfying the brain's needs for these compounds — especially in formulating foods for people who might already have some disturbance involving catecholaminergic or cholinergic neurons (e.g., the aged). The fact that tyrosine may not be an essential amino acid, nor choline an essential growth factor, for the young rat should not obscure our recognition that these compounds, like TRP, are absolutely essential for a normally functioning nervous system.

Acknowledgements. These studies were supported in part by grants from the United States National Institutes of Health, The National Aeronautics and Space Administration, and the Center for Brain Sciences and Metabolism Charitable Trust.

References

Blusztajn JK, Wurtman RJ (1982) Biosynthesis of choline by a preparation enriched in synaptosomes from rat brain. Nature 290: 417–418

Blusztajn JK, Wurtman RJ (1983) Choline and cholinergic neurons. Science 221: 614–620

Fernstrom JD, Wurtman RJ (1971) Brain serotonin content: increase following ingestion of carbohydrate diet. Science 174: 1023–1025

Fernstrom JD, Wurtman RJ (1972) Brain serotonin content: physiological regulation by plasma neutral amino acids. Science 178: 414–416

Fernstrom JD, Wurtman RJ, Hammarstrom-Wiklund B, Rand WM, Munro HN, Davidson CS (1979) Diurnal variations in plasma concentrations of tryptophan, tyrosine and other neutral amino acids: effect of dietary protein. Am J Clin Nutr 32: 1912–1922

Heraief E, Burckhardt P, Mauron C, Wurtman J, Wurtman RJ (1983) The treatment of obesity by carbohydrate deprivation suppresses plasma tryptophan and its ratio to other large neutral amino acids. J Neural Transm 57: 187–195

Hirsch MJ, Growdon JH, Wurtman RJ (1978) Relations between dietary choline intake, serum choline levels, and various metabolic indices. Metabolism 27: 953–960

Levy R, Little A, Chuaqui P, Reith M (1983) Early results from double-blind, placebo-controlled trial of high dose phosphatidylcholine in Alzheimer's disease. Lancet 1: 987–988

Madras BK, Cohen EL, Messing R, Munro HN, Wurtman RJ (1974) Relevance of serum-free tryptophan to tissue tryptophan concentrations. Metabolism 23: 1107–1116

Mauron C, Wurtman RJ (1982) Co-administering tyrosine with glucose potentiates its effect on brain tyrosine levels. J Neural Transm 55: 317–321

Pardridge WM (1977) Regulation of amino acid availability to the brain. In: Wurtman RJ, Wurtman JJ (eds) Nutrition and the brain. Raven, New York, pp 141–204

Spring B, Maller O, Wurtman J, Digman L (1983) Effects of protein and carbohydrate meals on mood and performance. J Psychiatr Res 17: 155–167

Wurtman JJ (1983) The carbohydrate-craver's diet. Houghton-Mifflin, Boston

Wurtman JJ, Wurtman RJ (1979) Drugs that enhance central serotoninergic transmission diminish elective carbohydrate consumption by rats. Life Sci 24: 895–904

Wurtman JJ, Wurtman RJ, Growdon JH, Henry P, Lipscomb A, Zeisel S (1981) Carbohydrate craving in obese people: suppression by treatments affecting serotoninergic transmission. Int J Eating Dis 1: 2–15

Wurtman JJ, Moses PL, Wurtman RJ (1982) Prior carbohydrate consumption affects the amount of carbohydrate that rats choose to eat. J Nutr 113: 70–78

Wurtman RJ (1982) Nutrients that modify brain function. Sci Am 246: 42–51

Wurtman RJ (1983) Behavioural effects of nutrients. Lancet 1: 1145–1147

Wurtman RJ, Hefti F, Melamed E (1980) Precursor control of neurotransmitter synthesis. Pharmacol Rev 32: 315–335

Impaired Control of Appetite for Carbohydrates in Some Patients with Eating Disorders: Treatment with Pharmacologic Agents

J. J. Wurtman and R. J. Wurtman[1]

The ability of animals or humans to control their consumption of energy in proportion to energy use has been recognized for some time (Kissileff and Van Itallie 1982). Less attention has been given to their ability to regulate intake of specific macronutrients. Although the roles of protein, carbohydrate, and fat in the body have been described and the amount of these nutrients needed daily has been established, consumption of foods has largely been regarded as motivated solely by the individual's need for energy. For example, if an animal given rat chow diluted 50% with fat (so that it now contains only half as much protein or carbohydrate per gram), elects to eat abnormally large amounts of this food, the overconsumption is generally interpreted as reflecting an inability to control energy intake in the presence of a high-fat diet (Kissileff and Van Itallie 1982). The possibility that the animal is eating more of the food in order to consume a desired amount of protein or carbohydrate by consuming quantities large enough to compensate for the dilution of one or the other nutrient is usually not considered.

This lack of attention to the regulation of macronutrient intake (as opposed simply to energy intake) is due in part to the way that feeding experiments measuring hunger and satiety tend to be designed. Traditionally such experiments are carried out by manipulating the animal's hunger and then measuring its consumption of a single test diet containing macronutrients in a fixed ratio. What usually is measured is the animal's hunger for calories, not for *specific* macronutrients, since the animal is unable to modify its intake of, for example, protein without also modifying proportionately its intakes of carbohydrates and fat. If an experimental treatment has changed the animal's hunger for a specific macronutrient such as protein, there is no way that the animal can demonstrate this change.

An experimental paradigm developed by Musten et al. (1974) now makes it possible to distinguish between an animal's decision to consume food for that food's caloric content and the decision to consume the food in order to obtain a particular nutrient in the food. Animals are allowed to choose their foods from a pair of diets whose calorie contents are identical but which contain different amounts of the macronutrient being examined. Using this technique, Musten et al. showed that rats can regulate their consumption of protein independently of their consumption of calories, and we demonstrated that rats can also regulate their carbohydrate intake independently of calorie intake (Wurtman and Wurtman 1979a, b). In our studies, animals are presented with a pair of

---

1 Massachusetts Institute of Technology, Department of Nutrition and Food Science, Cambridge, MA 02139, USA

isocaloric diets containing differing proportions of carbohydrate but equal amounts of protein. Our interest in studying carbohydrate intake stemmed from earlier studies showing that the synthesis and release of brain serotonin are dependent on the relative proportions of carbohydrate and protein consumed in the previous meal (Fernstrom and Wurtman 1972; Fernstrom et al. 1979). It seemed possible that the ingestion of carbohydrate is not random and is not motivated by the animal's energy needs, but reflects independent regulation mediated in part through brain serotonin.

In a typical experiment, we give rats a pair of isocaloric, isoprotein diets containing 75% or 25% carbohydrate. Daily food and carbohydrate intakes are measured for 3 weeks. The rats tend to consume a constant proportion of their daily total food intake as carbohydrate, i.e., about 60%–65%, depending on age and sex (Wurtman and Wurtman 1979a, b). This percentage was unchanged when we substituted a pair of diets containing 50% and 75% carbohydrate (Wurtman and Wurtman 1979a, b), indicating that carbohydrate consumption is indeed regulated separately from calorie intake.

This first set of experiments used diet pairs containing a carbohydrate of moderate sweetness (dextrose). Since many investigators have stressed the importance of taste in regulating carbohydrate consumption (for review, see Pfaffman 1977), we also examined the regulation of several carbohydrates of varying tastes. Rats were given access to pairs of isocaloric, isoprotein diets containing 25% or 75% sucrose, or dextrose or dextrin. The tastes of these diet pairs ranged from very sweet to bland. If taste was indeed an important influence on carbohydrate intake, we anticipated that animals would choose a higher proportion of sucrose when given the sucrose diet pair than of dextrose or dextrin when given these diet pairs. However, we found that the animals consumed the same proportion of their total food intake as carbohydrate regardless of the sweetness of the test diets (Wurtman and Wurtman 1979a, b).

The role of brain serotonin in regulating carbohydrate consumption was studied by treating rats with drugs known to enhance serotoninergic neurotransmission (e.g., dl-fenfluramine, which releases the transmitter into synapses) and measuring subsequent carbohydrate consumption. This treatment caused animals to eat significantly less of the 75% carbohydrate diet, without changing the amounts of the 25% carbohydrate diet that they consumed (Wurtman and Wurtman 1979a, b).

These results indicated that brain serotonin might be involved in a complex behavioral feedback loop in which the consumption of a high-carbohydrate meal enhances serotonin-mediated neurotransmission which, in turn, causes the animal to decrease its intake of carbohydrate during the next meal. This hypothesis was tested by examining food choice in fasting animals allowed to consume a small (6 calories) premeal containing either carbohydrate alone or fat plus carbohydrate and protein. Ninety minutes later, the animals were allowed access to a pair of isocaloric, isoprotein, 25% and 75% carbohydrate diets. Those eating the carbohydrate premeal subsequently consumed significantly less of the high-carbohydrate diet but the same total number of calories as those eating the mixed premeal (Wurtman et al. 1983), confirming that carbohydrate intake is regulated independently of calorie intake.

We extended our studies on the regulation of carbohydrate intake to human subjects, specifically those who claimed to have a well-recognized urge to eat carbohydrate foods either at meals or as snacks (J. J. Wurtman 1983). A pilot study was done with volunteers of normal weight whose claims of carbohydrate craving were evaluated through interviews and questionnaires. (In this and the subsequent human studies to be described, all volunteers signed consent forms that had been formally approved by the MIT Committee on Use of Humans as Experimental Subjects and by the Clinical Research Center Committee, which oversees studies on human volunteers.)

Eleven subjects were studied as outpatients. They were first asked to keep a record of their meal and snack food intake for 8 days. Most of the subjects consumed 60% or more of their snacks as high-carbohydrate foods, and each tended to eat these snacks during a period of the day or evening that was characteristic of him or her (Wurtman and Wurtman 1981). The subjects were then asked to restrict their carbohydrate snack intake to a prechosen high-carbohydrate food (such as cookies) and to eat only this snack food during the 4-h period when their cravings tended to be most intense. The subjects received tryptophan (2 g), dl-fenfluramine (20 mg), or placebos for these compounds in a double-blind crossover schedule. Each treatment period lasted 5 days and was followed by a 5-day washout period. The subjects were instructed to take their designated pills 1 h before the onset of their carbohydrate snack period, and then to record the number of snacks eaten during the test period. Fenfluramine significantly reduced the number of carbohydrate snacks consumed by the group as a whole; tryptophan did not have a significant effect within the group but did reduce snack intake in three of the subjects (Wurtman and Wurtman 1981).

This pilot study revealed that some individuals have a definite "need" to snack on high-carbohydrate foods at a particular time of day, and that this appetite for carbohydrates can be reduced following a treatment enhancing serotonin-mediated neurotransmission. Since anecdotal reports from the obese had suggested an inability to control carbohydrate consumption, and since this inability was often proposed as the reason for failure to lose weight or maintain weight loss, we decided next to study a group of obese carbohydrate cravers. We were especially interested in determining whether the overconsumption of carbohydrate by obese individuals was due, as had been suggested (Le Magnen 1967), to a failure to respond normally to the sweet taste of carbohydrate. [Le Magnen had proposed that changes in an animal's energy needs normally cause it to consume or reject sweet carbohydrates, and that this mechanism failed to operate among the obese. Obese adults and adolescents also reportedly are less sensitive to the taste of sucrose solutions, and improve their sensitivity with weight reduction (Grinker et al. 1976).]

Our second clinical study (Wurtman et al. 1981) was designed to see whether obese adults who considered themselves carbohydrate cravers would indeed snack preferentially on high-carbohydrate sweet or non-sweet foods if allowed 24-h access to an assortment of isocaloric carbohydrate-rich and protein-rich foods. If the overeating of carbohydrate was due to an altered taste sensibility, then sweet carbohydrate snacks would probably be chosen. And if obese individuals suffered from a general inability to control food intake, then both the protein-rich and the carbohydrate-rich snack foods would be consumed indiscriminately. However, if the carbohydrate hunger reported by these people was specific and if it was independent of the sweetness of foods, then we might observe snacking on any of the high-carbohydrate snack foods.

The study was conducted over a 4-week period. Subjects lived in an MIT dormitory; they were free to leave the facility but not the campus. Subjects had no choice of foods at meals; the menus were predetermined to meet daily nutrient needs. The total daily caloric content of the meals ranged from 950 to 1,000 calories. Since this was below the usual calorie intake of the subjects they were encouraged to snack freely from 10 isocaloric protein-rich and carbohydrate-rich snack foods (i.e., potato chips; bagel and cream cheese; cookies; chocolate candies; a chocolate or cranberry muffin; ham and cheese; meatballs; salami and cheese; barbecued pork chops; and miniature frankfurters). The snacks were dispensed in a refrigerated vending machine operated by a microcomputer. Subjects gained access to the snacks by typing a personal code that opened up all the doors of the vending machine. Once a snack was removed, the computer recorded its calorie, protein or carbohydrate content, the name of the subject who took it, and the time of day.

Subjects received no treatment during the first 2 weeks of the study. During the second 2-week period, one-third of the subjects received dl-fenfluramine three times daily in doses of 15–20 mg (the dose was reduced during the latter portion of the study to diminish the side-effect of sleepiness). Another third received L-tryptophan (800 mg three times daily), and the last third received placebo.

Twenty-three subjects completed the study. The group indeed did consume significantly more carbohydrate-rich than protein-rich snacks per day (4.1 ± 0.4 vs 0.8 ± 0.3). There was no consistent pattern in the selection of sweet or starchy snacks; indeed the sweet chocolate cupcake initially provided was replaced by a cranberry muffin because the chocolate cupcake was found to be "too sweet". Seventeen subjects failed to consume any protein-rich snacks at all during the study (Table 1). Each subject consumed a constant number of carbohydrate-rich snacks daily and at a characteristic time of day. Fenfluramine reduced carbohydrate intake significantly in the group as a whole (Table 2), while tryptophan decreased carbohydrate snack intake among three of the six subjects in its treatment groups (Table 3). These results confirmed our earlier findings with subjects of normal weight: a specific appetite for carbohydrate in individuals can be demonstrated, which is likely to be most pronounced at a time of day that is characteristic for each person and can be suppressed by treatments that increase serotonin release within the brain.

Neither study was designed to monitor carbohydrate and protein consumption at meals: thus we had no information on the proportions of carbohydrate to protein that carbohydrate-cravers consumed at meals or on possible effects of fenfluramine on total (meal plus snack) protein and carbohydrate consumption. Thus a third inpatient study was designed to allow subjects to chose from a variety of isocaloric protein and carbohydrate foods at mealtimes as well as at snacktimes.

Table 1. Effect of placebo on carbohydrate snack consumption (Wurtman et al. 1981)

Subjects	Carbohydrate snacks/day	
	Control	Placebo
A. L.	3.0 ± 0.31	2.7 ± 0.39
F. B.	1.6 ± 0.33	1.4 ± 0.42
E. N.	3.2 ± 0.43	4.0 ± 0.28
J. R.	5.3 ± 0.27	7.4 ± 0.42
G. N.	3.0 ± 0.56	2.7 ± 0.47
D. T.	2.7 ± 0.57	2.7 ± 0.48
B. C.	2.3 ± 0.39	1.8 ± 0.18

Subjects A. L., F. B., E. N., and J. R. (study 1) received no treatment for the first 2 weeks, and placebo (lactose 2.3 g/day in three divided doses) during the second 2 weeks. Subjects G. N., D. T., and B. C. (study 2) received placebo during both the first and second 2-week periods. In all cases, subjects did not know whether pills given during either period would be placebo or treatment. Comparisons for each subject were made between mean snack intakes during days 4 or 5 through 13, and 15 through 25 or 26, respectively. Data are expressed as means ± SEM

Table 2. Effect of fenfluramine on carbohydrate snack consumption (Wurtman et al. 1981)

Subjects	Carbohydrate snacks/day	
	Control	Fenfluramine
K. L.	7.7 ± 0.42	5.7 ± 0.37 (n.s.)
D. B.	1.0 ± 0.02	0.1 ± 0.08 ($P < 0.001$)[a]
P. A.	2.3 ± 0.33	1.8 ± 0.40 (n.s.)
B. D.	4.7 ± 0.53	1.1 ± 0.21 ($P < 0.001$)[a]
Ja. La.	4.6 ± 0.90	2.3 ± 0.47 (n.s.)
C. S.	2.7 ± 0.27	0.8 ± 0.21 ($P < 0.001$)[a]
Mi. Ri.	4.7 ± 0.42	3.1 ± 0.32 ($P < 0.001$)[a]
Ja. Li.	5.1 ± 0.70	3.5 ± 0.43 ($P < 0.05$)[a]
D. D.	5.1 ± 0.63	3.0 ± 0.66 ($P < 0.05$)[a]

Subjects K. L., D. B., P. A., B. D., Ja. La., and C. S. (study 1) received no treatment for the first 2-week period and fenfluramine (60 mg/day in three divided doses) during the second 2 weeks. P. A. received fenfluramine for 6 days and was then switched to placebo for 7 days; during this period she ate 2.8 ± 0.54 carbohydrate snacks/day. Subjects Mi. Ri., Ja. Li., and D. D. (study 2) received placebo for the first 2 weeks and fenfluramine (45 mg/day in three divided doses) for the second 2 weeks. Comparisons for each subject (except for P. A.) were made between mean snack intakes during days 4 or 5 through 13, and 15 through 25 or 26, respectively. Data are expressed as means ± SEM

[a] Significant difference from control group

Subjects were obese individuals who described themselves as carbohydrate cravers in questionnaires and interviews. They were studied as inpatients at the MIT Clinical Research Center. Subjects spent 2 days a week for 4 consecutive weeks at the Center. They were admitted on the evening before the first baseline day of the study and food intake measurements were made over the subsequent 48 h.

Table 3. Effect of tryptophan on carbohydrate snack consumption (Wurtman et al. 1981)

Subjects	Carbohydrate snacks/day	
	Control	Tryptophan
D. K.	10.5 ± 1.10	13.0 ± 0.63 (n.s.)
G. H.	4.7 ± 0.63	7.5 ± 0.67 ($P < 0.01$)[a]
K. N.	4.0 ± 0.53	2.2 ± 0.35 ($P < 0.02$)
D. C.	2.4 ± 0.43	1.3 ± 0.22 (n.s.)
M. B.	3.1 ± 1.10	3.6 ± 1.00 (n.s.)
Ma. Ri.	4.4 ± 0.33	2.6 ± 0.29 ($P < 0.001$)
P. R.	4.0 ± 0.31	3.6 ± 0.28 (n.s.)
L. H.	6.6 ± 0.36	4.7 ± 0.34 ($P < 0.05$)

Subjects D. K., G. H., K. N., D. C., and M. B. (study 1) received no treatment for the first 2 weeks and tryptophan (2.4 g/day in three divided doses) during the second 2 weeks. Subjects Ma. Ri., P. R., and L. H. (study 2) received placebo during the first 2-week period and tryptophan during the second 2 weeks (2.4 g/day in three divided doses). Comparisons for each subject were made between mean snack intakes during days 4 or 5 through 13, and 15 through 25 or 26, respectively. Data are expressed as means ± SEM
[a] Significant increase over control value

D-Fenfluramine (15 mg at 7 a.m. and at 4 p.m.) or its placebo was administered for 8 days in a double-blind crossover schedule. Food intake was measured on days 1, 7, and 8 of each treatment period. Subjects were allowed to go home during days 2–6.

Meals were taken in the Center's dining room and snacks were dispensed as in the second study, from a computerized vending machine near the patients' rooms.

The meals contained an assortment of isocaloric high-protein and high-carbohydrate foods typical of the foods eaten at the respective meals. For example, breakfast choices included scrambled eggs, Canadian bacon, cottage cheese, English muffins, pancakes with syrup, and granola cereal. Each dish provided 15–16 mg protein or carbohydrate; fat was added to make the choices isocaloric. (We have no information as to whether or not appetite for fat is also regulated.) The choice of foods provided at each meal remained the same throughout the study to minimize possible effects of novelty. (Parenthetically, the method of assessing nutrient intake used in this study may prove useful for examining other aspects of eating behavior, since it utilizes foods commonly eaten by the subjects and mimics, to some extent, the eating environment of a cafeteria. This method has an advantage over studies which use blenderized foods, liquid diets, or synthetic foods since it does not require that the subjects adjust to an unfamiliar eating situation.)

In none of these human studies did we see evidence of completely uncontrolled carbohydrate consumption. Rather, each subject tended to restrict his or her snacking to a characteristic time of day (Fig. 1) and to consume a characteristic number of snacks per day. This control of carbohydrate intake was not related to a desire to lose weight, since weight loss was forbidden during the study and subjects were expelled from it if we suspected that they were, in fact, dieting.

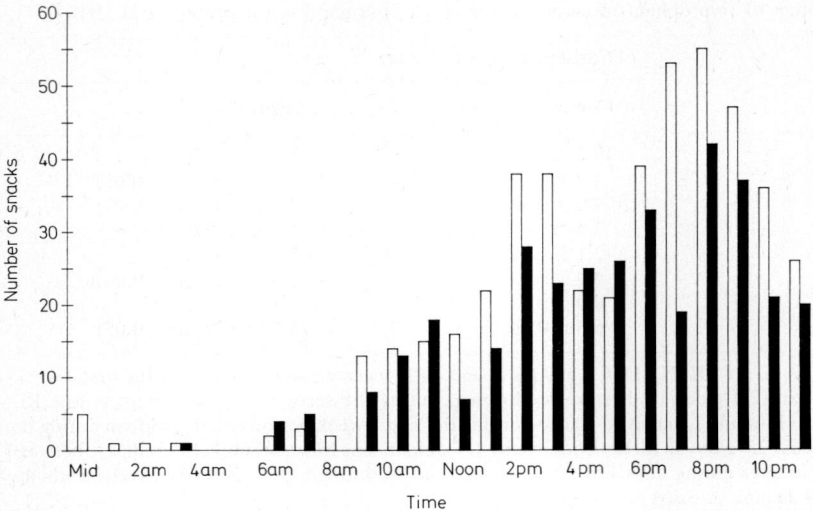

Fig. 1. Total number of carbohydrate snacks taken during each hour of the day and night by the 20 subjects during 3 placebo days *(open bars)* and 3 days of D-fenfluramine treatment (15 mg PO twice daily; *closed bars*). Each *bar* represents snacks consumed during the hour ending with the time indicated. Subjects had breakfast at 08:00–08:30 h; lunch at 12:15–12:45 h; and dinner at 17:00–17:30 h. The subjects were not allowed to snack during these meal intervals. Each subject consumed an average of 5.7 carbohydrate-rich snacks and 0.7 protein-rich snacks per day during the placebo period. All snacks contained 105–110 kcal; the carbohydrate snacks contained 10–12 g carbohydrate

Yet "bingeing" is a commonly reported problem among the obese, and their failure to maintain weight loss is often anecdotally attributed to an inability to limit the consumption of carbohydrate foods once the individual resumes eating them at all. These people often report that eating even a small amount of carbohydrate almost invariably leads to excessive carbohydrate consumption. Unlike the subjects in our studies, these individuals do not "shut-off" their hunger for carbohydrate until they consume massive quantities. Many individuals who report this behavior have attempted to lose weight by following a ketogenic diet that is very low in carbohydrates. (This conclusion is drawn from interviews of patients attending a weight loss clinic that used a protein-sparing diet and also from patients who came to J. W. for weight-loss counseling.)

Thus we carried out an animal study to determine whether the low-carbohydrate ketogenic diet might itself predispose to enhanced desire for, and consumption of carbohydrates (i.e., when the animals are given a choice). Rats were placed on a carbohydrate-free, ketogenic diet containing 30% protein, 37% fat, and 33% cellulose. Control rats were given the standard carbohydrate-containing diet (65% carbohydrate). Ketosis was induced in the test rats after 2–3 weeks with only a negligible loss in weight (Wurtman et al. 1983). Carbohydrate intake by the ketotic and control rats was tested by allowing both groups access to a pair of isocaloric, isoprotein, 25% and 75% dextrin diets. The ketotic rats ate significantly more carbohydrate than the controls; the same results were obtained when sucrose was substituted for dextrin in the diet pairs. Thus the ketotic rats showed the same inability to control carbohydrate intake as that reported by individuals on low-carbohydrate diets.

This excessive consumption of carbohydrate may have been caused, in part, by the well-known decrease in insulin secretion that follows carbohydrate deprivation (Carmel et al. 1979; Schemmel et al. 1982; Unger et al. 1963). Tryptophan, the precursor of serotonin, depends for its entry into the brain on an insulin-mediated decrease in the plasma levels of five other neutral amino acids that compete with tryptophan for transport across the blood-brain barrier (i.e., leucine, isoleucine, valine, tyrosine, phenylalanine). If brain levels of tryptophan were lower than normal because of a decrease in postprandial insulin secretion, then insufficient serotonin might be synthesized and released to regulate carbohydrate consumption normally. Thus we measured brain tryptophan in carbohydrate-deprived animals given brief access to a carbohydrate-containing food. All rats were fasted for 4 h starting at the onset of the dark period. After the fasting period, animals from the control and carbohydrate-deprived groups were killed and their brains assayed for tryptophan. Subsequently the remaining control rats were allowed to eat 6 g/kg of a carbohydrate-rich food. Half the remaining carbohydrate-deprived animals were given the same amount of carbohydrate (6 g/kg, carbohydrate-restricted group) and the others were allowed to consume carbohydrate ad libitum (they ate 16 g/kg). Ninety minutes later, animals were killed and their brains were assayed for tryptophan.

Brain tryptophan levels were lower in fasting than in fed animals (Table 4), but increased significantly after ad libitum feeding in control animals (which ate 6 g carbohydrate) and in previously deprived animals (which ate 16 g). The increase in tryptophan in the latter group was no greater than in control animals, despite their consumption of more than twice as much carbohydrate (Table 4). Moreover, the previously deprived group which subsequently was pair-fed with control rats, thus consuming 6 g carbohydrate, failed to show a significant increase in brain tryptophan. These results indicate that after prolonged carbohydrate deprivation the consumption of dietary carbohydrate is not followed by the usual increase in brain tryptophan. The increase in tryptophan (and also, presumably in serotoninergic transmission) is achieved only after the carbohydrate-deprived animals consume excessive quantities of carbohydrate.

The laboratory and clinical studies described above point to the existence of a specific appetite for carbohydrate-rich foods and to the involvement of brain serotonin in its regulation. When an animal consumes a meal rich in carbohydrate, its subsequent meal is chosen to contain proportionately more protein. However, prolonged carbohydrate deprivation may suppress this response. If similar processes occur in obese people, they could contribute to "carbohydrate bingeing" — a problem that might be exacerbated by the insulin insensitivity characteristic of obesity (Heraief et al. 1983).

Even though carbohydrate craving may be *exacerbated* by prolonged consumption of an inappropriate diet (i.e., one that tends to decrease brain tryptophan and serotonin levels), it is still necessary to explain why some people

Table 4. Effects of carbohydrate intake or deprivation on brain tryptophan

Experimental group		Corbohydrate consumed (g/kg)	Tryptophan (µg/g)
Control	Fasting	–	3.8 ± 0.13*
Control	Ad lib-fed	6	4.8 ± 0.13
Carbohydrate-deprived	Fasting	–	2.6 ± 0.20*
Carbohydrate-deprived	Ad lib-fed	16	4.9 ± 0.65
Carbohydrate-deprived	Pair-fed	6	3.3 ± 0.32*

Adult male rats consumed either a control (+, 60% CHO, 30% protein, 10% fat) or a carbohydrate-deficient (++, 0% CHO, 30% protein, 37% fat, 33% cellulose) diet for 4–6 days, during which time all of the latter group became ketotic. On the day of the experiment, all animals were fasted for 4 h at the start of the dark period. Thereafter the control animals and half the carbohydrate-deprived animals (pair-fed) were given 6 g/kg of carbohydrate, an amount corresponding to what control animals usually consume in the first 30 min of eating. The remainder of the carbohydrate-deprived group was given 6 g/kg of carbohydrate (ad lib-fed), corresponding to what this group normally consumes in the 30 min. Ninety minutes after the introduction of the food all rats were decapitated and brains were assayed for tryptophan

Values are expressed as means and SEM. Statistical analyses were done using one-way ANOVA and the Tukey's test. * Significant difference ($P < 0.05$) from corresponding ad lib-fed group

tend to snack excessively on carbohydrates while others do not. (It also should be recognized that the propensity to snack on carbohydrates need not necessarily lead to obesity: many such people compensate for it by reducing mealtime calorie intake.) We observed in our most recent study on the effects of fenfluramine on daily food intake that D-fenfluramine decreases *snack* consumption of carbohydrates (and calories) to a significantly greater extent than it diminishes *mealtime* carbohydrate intake (unpublished observations). This suggests that different mechanisms underlie these types of behavior. Perhaps the desire to consume a certain quantity of carbohydrate at meals, or rather a certain proportion of carbohydrate to protein, reflects the normal operation of a brain mechanism concerned with maintaining good nutrition, while the desire to consume purely carbohydrate snacks in the middle of the afternoon or evening reflects the brain's "desire" for serotonin, unrelated to the body's nutritional needs. Serotonin release is known to increase drowsiness, to facilitate sleep onset, and to diminish pain sensitivity (R. J. Wurtman 1983). Moreover, most antidepressant drugs share with dietary carbohydrates the propensity to enhance serotonin-mediated neurotransmission (either by blocking serotonin's intracellular metabolism by monoamine oxidase or by suppressing its reuptake into the presynaptic terminals that release it). Perhaps people with obesity or other appetitive disorders related to inappropriate carbohydrate consumption are, in reality, consuming the snacks for their psychopharmacologic effects.

Acknowledgements. These studies were supported in part by grant from the National Institutes of Health (AM-14228) and the National Aeronautics and Space Administration (NAG-2-210).

References

Carmel N, Koijn A, Kaufman N, Guggenheim K (1979) Effects of carbohydrate-free diets on the insulin-carbohydrate relationships in rats. J Nutr 105: 1141–1149

Fernstrom JD, Wurtman RJ (1972) Brain serotonin content: physiological regulation by plasma neutral amino acids. Science 178: 414–416

Fernstrom JD, Wurtman RJ, Hammarstrom-Wiklund B, Rand WM, Munro HN, Davidson CS (1979) Diurnal variations in plasma concentrations of tryptophan, tyrosine, and other neutral amino acids: effect of dietary protein intake. Am J Clin Nutr 32: 1912–1922

Grinker J, Price J, Greenwood M (1976) Studies of taste in childhood obesity. In: Novin D, Wyrwicka W, Bray G (eds) Hunger, basic mechanisms and clinical implications. Raven, New York, pp 441–457

Heraief E, Burckhardt P, Mauron C, Wurtman J, Wurtman R (1983) Obesity and its dietary treatment may suppress synthesis of brain serotonin. J Neural Transm 57: 187–195

Kissileff H, Van Itallie T (1982) Physiology of the control of food intake. Annu Rev Nutr 2: 371–418.

Le Magnen J (1967) Habits and food intake. In: Code CF (ed) The alimentary canal. American Physiological Society, Washington DC, pp 11–30 (Handbook of physiology, section 6)

Musten B, Peace D, Anderson GH (1974) Food intake regulation in the weanling rat: self-selection of protein and energy. J Nutr 104: 563–572

Pfaffman C (1977) Biological and behavioral substrates of the sweet tooth. In: Weiffenback JM (ed) Taste and development: the genesis of sweet preference. US Government Printing Office, Washington DC, pp 3–24

Schemmel R, Hu D, Mickelson O, Romsos D (1982) Dietary obesity in rats: influence on carbohydrate metabolism. J Nutr 112: 223–230

Unger R, Eisentraut A, Madison L (1963) The effects of total starvation upon the levels of circulating glucagon and insulin in man. J Clin Invest 42: 1031–1039

Wurtman JJ (1983) The carbohydrate craver's diet. Chapter 1: Carbohydrate craving. Houghton-Mifflin, Boston, pp 1–15

Wurtman JJ, Wurtman RJ (1979a) Fenfluramine and other serotoninergic drugs depress food intake and carbohydrate consumption while sparing protein consumption. Curr Med Res Opin 6: 28–33

Wurtman JJ, Wurtman RJ (1979b) Drugs that enhance central serotoninergic transmission diminish elective carbohydrate consumption by rats. Life Sci 24: 895–904

Wurtman JJ, Wurtman RJ (1981) Suppression of carbohydrate consumption as snacks and at mealtime by dl-fenfluramine or tryptophan. In: Garrattini S (ed) Anorectic agents: mechanisms of actions and of tolerance. Raven, New York, pp 169–182

Wurtman JJ, Wurtman RJ, Growdon JH, Henry P, Lipscomb A, Zeisel S (1981) Carbohydrate craving in obese people: suppression by treatments affecting serotoninergic transmission. Int J Eating Dis 1: 2–11

Wurtman JJ, Moses PL, Wurtman RJ (1983) Prior carbohydrate consumption affects the amount of carbohydrate that rats choose to eat. J Nutr 113: 70–78

Wurtman RJ (1983) Behavioral effects of nutrients. Lancet 1: 1145–1147

Animal Models: Anorexia Yes, Nervosa No

N. Mrosovsky[1]

In 1982 a case history appeared of a 13-year-old girl fulfilling the DSM-III criteria for anorexia nervosa (Weller and Weller 1982). The patient was obsessed about food and being fat; the diagnosis was confirmed by four psychiatrists. The unusual point about this case was that there was also a tumor infiltrating one side of the hypothalamus. This is probably the first case where characteristics of cachexia associated with a brain tumor closely resemble those common in anorexia nervosa. "It would seem probable", the authors say, "that the tumor and the clinical syndrome were related." This conclusion is highly debatable. First, there are thousands of cases of anorexia nervosa where no tumors have been found. The occasional combination of the two disorders may be a chance event. It is necessary to know the incidence of hypothalamic tumors and anorexia nervosa in the population before making valid conclusions. Second, although the anorexia abated after irradiation, CT scans showed no change in the lesions. Third, unilateral lesions would be unlikely to lead to severe weight loss. Unilateral lateral hypothalamic (LH) lesions, made deliberately in people in attempts to control obesity, failed to achieve this (Quaade et al. 1974).

The Wellers' paper is a recent and vivid example of an idea recurrent in the literature on anorexia nervosa, that the disorder stems from hypothalamic dysfunction. And since there are a number of animal preparations with dramatically altered feeding patterns and with hypothalamic damage, this idea leads naturally to the question: "Are there animal models of anorexia nervosa?" If such models could be developed, they would offer many possibilities for furthering our understanding of the disease. Unfortunately there are numerous reasons why little hope should be pinned on this outcome.

First, there is no compelling evidence that hypothalamic dysfunction is primary in anorexia nervosa. Correlations between the percentage below ideal body weight and the delay in thyroid hormone peak after TRH injection, and the rate of core temperature change under heat or cold stress suggest rather that dysfunctions are secondary to weight loss (e.g., Vigersky and Loriaux 1977). The relationships of dysfunctions to body weight are by no means simple (Weiner 1982). For example, weight recovery does not always lead to normalization of luteinizing hormone levels; abnormal growth hormone dynamics have been noted in bulimia without emaciation (Mitchell and Bantle 1983). However, such observations do not conclusively dissociate dysfunction from weight loss, because ideal weights rather than the subject's own physiologically defended weights are used as reference levels. The two may differ greatly, especially when

---

[1] University of Toronto, Departments of Psychology and Zoology, Toronto M5S 1A1, Ontario, Canada

there has been a previous history of overweight, as is not uncommon in anorexia nervosa.

Second, there are prominent psychological aspects of the disorder. Even if one takes the extreme position that all of these are secondary to weight loss, such things as "implacable attitude towards eating, food or weight that overrides hunger . . . denial of illness . . . desired body image of extreme thinness" (Feighner et al. 1972) must at least play a major role in sustaining the disease. For anyone who takes these features seriously, the idea of animal models must border on the absurd.

Third, when specific options for animal models are examined they are found wanting. A number of candidate models will now be considered, starting with those concerned with the causes and primary aspects of the disorder and moving on to those concerned with the consequent secondary features.

Some Options for Animal Models of the Etiology of Anorexia Nervosa

Lateral Hypothalamic-Lesioned Animal

Biologists who have worked with LH animals have pointed out that the full features of the syndrome differ from those of anorexia nervosa. For example, LH rats are often adipsic, while anorectic patients drink more than normal, if anything. The main similarity is in the reduction of eating, but the aphagia of animals with large LH lesions may be a nonspecific effect consequent on sensorimotor impairments that depress a variety of behaviors (Stricker and Andersen 1980). With smaller LH lesions sensorimotor deficits are relatively unimportant; reduced intakes do not depend on an inability to eat but rather on a lowering of the defended level or set point for body weight. Until the new set point is attained the rats are anorectic, but in the clinical syndrome there is no true loss of appetite. Once the new set point is reached eating and energy expenditure of LH rats return to values appropriate to that body mass. In contrast, stability of weight is uncharacteristic of anorexia nervosa and hyperactivity often persists (Keesey, to be published).

Naturally Occurring Animal Anorexias

Natural anorexias are widespread in the animal kingdom at times when some more important activity conflicts with eating. For instance, in the incubating junglefowl, a species where the female sits on the eggs unattended and unfed by the male, there is a 10%–15% drop in body weight during the incubation period. This takes place even if food is put so close to the hen that she can reach it without leaving the nest. Presumably natural selection favors animals that guard and warm the eggs rather than leave them for foraging. The mechanism that has evolved to ensure nest attentiveness is a progressive lowering of the set point, not an abandonment of regulation altogether. This can be demonstrated by removing what little food the hen is eating and then returning it after a few days; the hen then eats more than controls and regains some of the weight lost before resuming weight loss (Mrosovsky and Sherry 1980). Such sliding set points for defended body weights have been found in mammalian hibernators in the course of their cycles of fat deposition and loss. They probably underlie the anorexias that occur in a large range of species (Mrosovsky and Sherry 1980).

Animal anorexias provide examples of natural changes in the regulated level of fat and for this reason are not good models of what happens in anorexia nervosa. In the clinical syndrome, except perhaps in terminal stages, appetite is present (Garfinkel and Garner 1982). Presumably appetite remains because the set point for body weight remains relatively high compared with the actual weight of the patient. That there is such a discrepancy between the set and the actual weight is inferred from increased efficiency of food utilization (Walker et al. 1979), attenuated negative alliesthesia (Garfinkel et al. 1978), preoccupation with food similar to that in starvation during famine or shipwreck, and some other possible indices. The data are reviewed more fully elsewhere (Mrosovsky 1983, to be published).

Pharmacological Blockade of Hypothalamic Noradrenergic System

Leibowitz (to be published) has discussed the possibility that some of the symptoms of anorexia nervosa may be attributable to a "basic primary hypothalamic disorder". This idea relates to the knowledge that paraventricular noradrenaline injections stimulate eating and weight gain. Pharmacological blockade of the α-noradrenergic system produces the opposite effects. So the proposal appears to be that anorexia nervosa is produced in part by a decrease in hypothalamic noradrenergic activity (other transmitters may also be involved). The problem here is that loss of appetite is not characteristic of anorexia nervosa, but when a rat with plenty of food nearby in its cage is given a noradrenergic blocker and then eats less, the most obvious inference is that its appetite has been affected. Poor appetite is probably dependent on a lowering of the defended level of body weight. At least this appears to be the case with some anorectic drugs that are known to affect CNS transmitters, amphetamine and fenfluramine for example. The evidence comes from experiments in which rats were reduced in body weight prior to administration of these drugs. If food was provided ad libitum at the same time as the drug treatment was begun the animals gained weight initially and ate more than the heavier drug group, demonstrating that ability to eat was not seriously impaired. However, weight did not return to that of controls but only to a level characteristic for that particular dose (Fig. 1). The design of the experiment is similar to that used by Powley and Keesey (1970) to demonstrate that LH lesions lower set points, and similar inferences can be drawn from the results. In contrast, as already

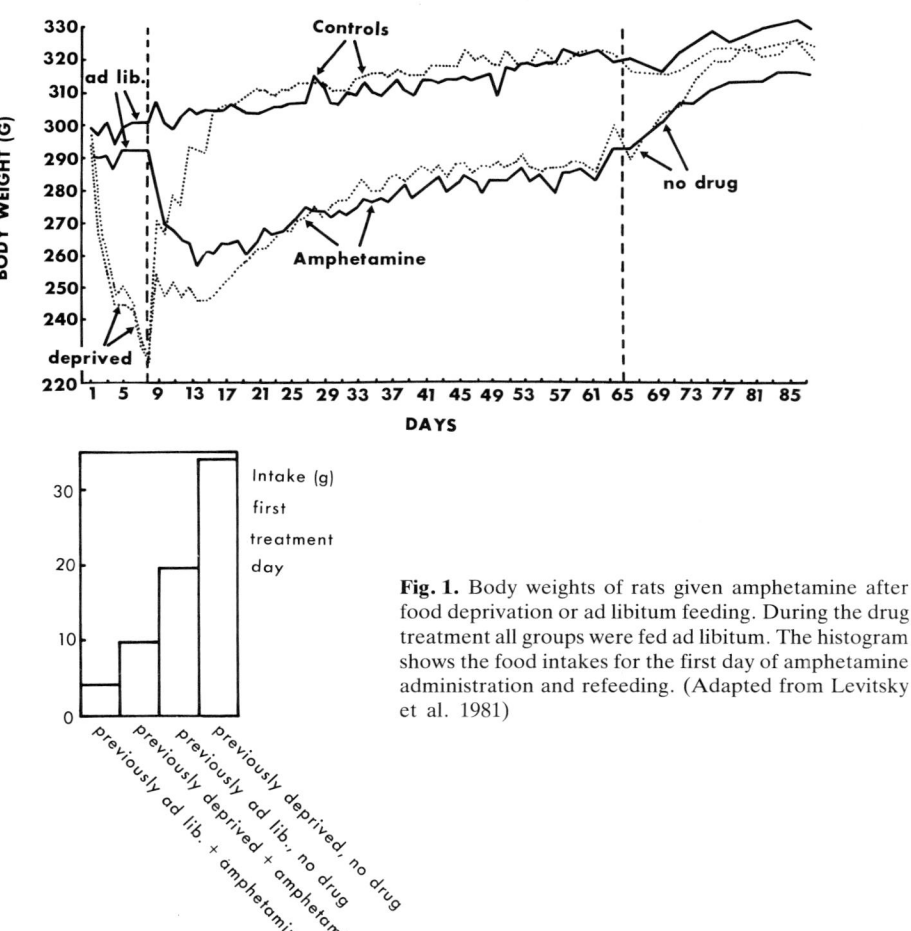

Fig. 1. Body weights of rats given amphetamine after food deprivation or ad libitum feeding. During the drug treatment all groups were fed ad libitum. The histogram shows the food intakes for the first day of amphetamine administration and refeeding. (Adapted from Levitsky et al. 1981)

mentioned, there is no evidence that set points are lowered in anorexia nervosa.

Leibowitz (to be published) has also proposed that increased noradrenergic activity could be responsible for bulimic bouts. She noted that food deprivation results in increased noradrenaline turnover in the paraventricular nucleus. The implication here is that excessive dieting in anorexia nervosa might stimulate the noradrenergic activity and so bring on a binge. The idea that bulimic bouts represent compensatory activity of an essentially undamaged regulatory system has probably occurred to many people (e.g., Mrosovsky, to be published; Coscina and Dixon 1983). But if this is what is happening, and if noradrenergic activity is the physiological basis of this reaction to lack of food, then these changes are secondary, not primary, in anorexia nervosa. Perhaps, all the same, it would be worthwhile to try to ascertain whether there are differences in noradrenergic functioning between anorectics displaying binges and those restricting more consistently. However, many substances affect feeding: insulin, CCK, glucagon, bombesin, estrogen, endorphins, etc. Concentration on hypothalamic noradrenaline may be an inspired guess, but at present it seems rather arbitrary.

"Self-Starvation" and Activity

Epling et al. (to be published) have proposed that "excessive activity may be part of a causal chain which can result in extreme weight loss and death". Their paper draws attention to the scarcity of full descriptive studies on compulsive exercising in anorexia nervosa. Certainly at present there are not enough data to enable one to say how often hyperactivity precedes weight loss and other symptoms. Nor in fact do Epling et al. assert that hyperactivity is the only factor involved; what is important is the interaction between the opportunity to engage in activity and eating schedules. The "self-starvation" phenomenon in rats is considered as supporting evidence. If rats are given food for only 1 h a day and are able to run in a wheel during the rest of the day but not this hour, then they eat less than rats without wheels on the same feeding schedules. If the experiment is prolonged the animals are liable to die. In this particular situation the activity is very important because rats without the wheel do much better. But it must be pointed out that, wheel or no wheel, rats have difficulty in coping with such restricted schedules. Since the experimenter, not the rat, imposes the 1-h feeding regimen, the term self-starvation is silly. To the extent that the phenomenon has a human counterpart, the patient herself adopts the role of the experimenter in setting the eating patterns. The animal analogy offers no insight into why she does this.

Future Animal Models

In the past, discussion of animal models for anorexia nervosa has centered around the feeding disturbance. But if anorexia nervosa represents a battle between cognitive aggressors and physiological defenders that are essentially intact and occasionally reassert themselves in the form of bulimia, craving for food, and increased efficiency of utilization, then this emphasis is misplaced. If one thinks animal models are worth pursuing, it may be better to try to devise schedules that reinforce weight loss and punish consummatory behavior (Mrosovsky, to be published). Of course there may still be predisposing factors in the regulatory systems. Perhaps defense mechanisms for body fat levels are inherently weaker in some people and more liable to be overwhelmed. Perhaps set points are more easily adjustable in certain people. Individual differences in the precision of regulation and the levels around which that regulation occurs are interesting in their own right and can be studied in animals, but it may be more relevant to anorexia nervosa to concentrate more on the physiological basis for alleviation of anxiety, response to stress, and reward mechanisms in general.

Moreover, predisposing factors need precipitating factors before there is illness. Could animals ever model the interactions between particular constitutions and family dynamics, feelings of inadequacy, and body image distortions? In this context responses to stress by animals in different positions in social hierarchies, or anecdotal accounts of dogs pining away when their masters die, hardly seem worth pursuing.

Given the unpromising prospect for animal models, why has interest in them persisted? Probably there are several reasons. One may be that conclusions about the inappropriateness of animal models for anorexia nervosa are based mainly on lack of supporting information; there are few facts that clearly dissociate the disease from the candidate models (Keesey, to be published; cf. McHugh and Moran 1977). The best evidence is the contrast in appetite and set points between anorexia nervosa and the animal preparations discussed above. However, some people are uneasy about inferring lack of appetite when an animal does not eat, even if motor and sensory capacities are checked; and they are reluctant to infer the presence of appetite in patients who are obsessed with food, yet eat very little. But with denial of illness being such a prominent feature of anorexia nervosa, less reliance should be placed on refusals to admit hunger than on admissions after recovery that hunger was experienced.

There are also those who find the concept of set point unacceptable without understanding its physiological basis. However, even without such knowledge, and even without making assumptions about the formal structure of the control systems, the term set point can be used descriptively in a definable way to refer to level around which weight regulation occurs (see Mrosovsky and Powley 1977, for terminology). Moreover, it is possible to quantify differences between actual weight and set-point weight. Mrosovsky (1983, to be published) has discussed several measures that can be applied to people. The available evidence suggests that anorectics are well below rather than close to their set points, but the data are sketchy and more research is needed. Keesey (to be published) has also called for independent measures of the relationship between actual and set weight, rather than reliance on deviations from ideal body weight.

Further work on this topic would be valuable for several reasons. It offers a chance of resolving whether or not anorectic patients are substantially below their set points, and so better defining the disorder — and at the same time purging the field of inappropriate animal models. More important, it could provide diagnostic tests for telling when a patient recovers not just to norms for her height but to her own physiologically defended level. For instance, it should be asked, initially with animal subjects, what is the threshold value of weight loss for the appearance of increased efficiency of food utilization. It should be asked, initially with healthy human subjects, what is the threshold of weight loss for the disappearance of negative alliesthesia. And in both cases it should be asked whether the response cuts in at full strength at threshold or grows thereafter in proportion to weight loss. The amount of key pecking and wheel running in chickens and rats, respectively, is proportional to the percentage body weight loss (Collier 1969), so it is likely that a proportional controller operates with other compensatory responses. There are already some indications that this is the case with utilization efficiency (Ozelci et al. 1978; Walker et al. 1979);

previously obese anorectics tend to recover weight more rapidly than do previously normal weight anorectics eating equivalent amounts (Stordy et al. 1977; Walker et al. 1979). Of course there will be individual variability in any indices developed for clinical use, but it should be possible to establish confidence limits for recovery to various percentage deviations from regulated levels of weight. This would help in tackling the question of whether the lack of clearer relationships between endocrine dysfunctions and body weight (Weiner 1982) results from reliance on ideal rather than set-point weight.

Secondary Changes

Animals may be inappropriate for modelling the primary features of the disease but they could be useful in unraveling possible secondary disorders. Study of spontaneous animal anorexias has shown that set points for body weight are plastic. Perhaps altered dietary habits in anorexia nervosa tap into this plasticity and so sustain the disorder (Mrosovsky 1983). The patient may put herself into a situation where she is

> longing still
> For that which longer nurseth the disease;
> Feeding on that which doth preserve the ill,
> The uncertain sickly appetite to please.
> (Shakespeare)

There are various possibilities.

Does Prolonged Dieting Lower Set Points?

The orthodox view, derived from the high rate of recidivism in people treated for obesity, has been that little can be done to lower the defended levels. But it has recently been pointed out that those seeking help from therapists are a selected population who have failed to control their weight by other means. If one goes out, as Schachter (1982) has done, and samples more widely in the community, one finds plenty of people who have by their own account successfully maintained reduced weights after dieting. Perhaps the gloomy view about the chances of maintaining weight loss is based on a study of the most refractory individuals.

Turning to work with animals, there are plenty of examples where previously deprived animals caught up to control levels on refeeding (e.g., Collier 1969; Bjorntorp and Yang 1982; Keesey, to be published), but there are also a number of cases where this did not happen (Szepesi 1980; Armstrong et al. 1980). Before it is profitable to argue about whether such results refute the idea of set points for body weight or suggest that such set points can be adjusted, it is necessary to address the question of how catch-up can be measured. For instance, Armstrong et al. (1980) fitted a straight line to a 23-day baseline and projected this out.

Weights of groups of rats deprived for a few days and then refed did not catch up to these projections in the subsequent 40 days. But there was no statistical treatment of whether the rats were significantly below the projected lines. Moreover, use of a straight line, however well it may fit a short section of data, is inappropriate for a growth curve that is known not to be a straight line over longer periods. Finally, body weight was only followed for 40 days; defense of body weight may be a prolonged process and perhaps the deprived groups did eventually catch up. In Szepesi's (1980) work also the rats were only followed for a relatively short period after refeeding. His experiment raises some interesting points, as although the deprived group was lower in weight than controls, the deprived animals ended up as fat (expressed as a percentage of body weight). These data make it clear that body weight is only a rough index of fatness, and further, that to address the question of catch-up properly, one needs to know what variable is regulated: the amount of fat, the percentage of fat in the body, the fat cell size, or something correlated with one of these. A convenient laboratory species that shows absolutely no growth as an adult would also be helpful.

One thing is established, however: it is very difficult, perhaps impossible, to reduce the number of fat cells by prolonged food restriction. Even when rats are deprived of all food for 1 week and then given glucose-electrolyte diet for the next 7 weeks there is no detectable loss of fat cells, even despite a 40% drop in body weight and virtual absence of lipid in adipocytes (Miller et al., to be published). On refeeding it took around 3 months for fat cell size to approach control levels, an example of the sluggishness of the compensatory mechanisms. If growing rats are restricted from 6 weeks to 6 months to 60% rations compared with ad libitum controls, then they have fewer fat cells (Bertrand and Masoro 1977). However, this may not be a true reduction in cell number but a diminution of the gradual proliferation that is a normal part of ontogeny (Faust et al. 1978). In other studies where the rats were returned to full rations at 6 months, fat cell number caught up to that of the controls, but this took months (E. J. Masoro 1983, personal communication).

Increased Efficiency of Food Utilization and Metabolic Memory

When animals are underweight after deprivation they use food more efficiently, gaining more body weight per gram of food ingested (Boyle et al. 1978; Levitsky et al. 1976; Ozelci et al. 1978; Bjorntorp and Yang 1982; Coscina and Dixon 1983). During recovery from anorexia nervosa too, the further patients are below their maximum weight before the disease, the fewer calories they need to gain 1 kg in weight (Walker et al. 1979). This phenomenon is often referred to as increased efficiency of food utilization. There has been considerable interest in the mechanisms of this increased efficiency. From a clinical point of view, whether it stems from a turning down of brown fat metabolism, an increase in lipogenic enzymes, or a combination of different causes is perhaps not as important as how long any "metabolic memory" (Szepesi 1973) lasts. In one experiment rats were given only 25% of the intake of freely feeding rats for 1 week and then pair-fed with these control rats for the next 6 weeks; increased gain in body weight and fat lasted for 3 weeks (Ozelci et al. 1978).

If deprivation history has lasting effects, it may make it harder for anorectics to re-establish normal eating habits (Coscina and Dixon 1983). Eating only a little may result in more gain than is tolerable, and this may reinforce fears of losing control and lead to further fasting, punctuated only perhaps by binges. Coscina and Dixon (1983) also report on an experiment with rats. After a 4-day deprivation rats put on more weight, especially when receiving a high-fat diet, than those that had never been deprived, and ended up at higher absolute levels. The authors speculate that the experience may have raised the body weight set point.

Much has to be clarified about these studies. First, the idea that deprivation can lead to a higher body weight set point is different from the idea that these metabolic adaptations contribute toward catch-up by increasing efficiency. The two may or may not be related. Second, set-point elevation needs checking with tests of the defense of the new weights, meticulously matched control groups, and experiments lasting longer. Third, metabolic adaptations in many cases are not firmly established. Such adaptations are not the same as increased efficiency of food utilization, which is defined in operational terms as the ratio of grams gained to grams eaten. When an animal is deprived and refed, this ratio is likely to alter for some not very exciting reasons. If the animal is hyperphagic when refed, then it is hardly surprising that it gains more weight than a control, because both require some calories for maintenance; calories ingested in excess of maintenance needs can be put into fat. Even if increased efficiency is found when previously deprived rats are then pair-fed amounts of food equal to those taken by controls it is not absolutely safe to infer metabolic adaptations, because the maintenance needs will be less for a rat that has a smaller mass after deprivation; it can therefore put more of the food into weight gain than the control. The purest experiments are those that eliminate increased efficiency resulting from hyperphagia or reduced maintenance needs by matching intakes between deprived and controls not in terms of actual amounts but as indexed to the body mass (Levitsky et al. 1976; Boyle et al. 1978; Ozelci et al. 1978). If this is not done there may be an overattribution of increased efficiency to metabolic adaptations. For example, the estimate of metabolic memory lasting some 3 weeks, cited above, might come down if the experiment was repeated with food intakes indexed to body weight. And the increased efficiency in the Coscina and Dixon experiment is surely partly attributable to reduced maintenance requirements.

Despite such caveats, these problems remain well worth studying. Set points are plastic and maybe deprivation is one way of changing them. Metabolic adaptations have been demonstrated. In the tests on anorectic patients, allowance was made for maintenance requirements and only calories in excess of these were used to calculate efficiency (Walker et al. 1979). What is needed now is to work out the details, how severe deprivations are needed, how persistent the effects are, and the influence of age and of repeated deprivations (Szepesi and Epstein 1977). Individual differences are obviously important but a further complication. Reactions of rats in deprivation and catch-up experiments are highly variable (Armstrong et al. 1980; Coscina and Dixon 1983). If only a subset of individuals is prone to lasting effects, there are methodological challenges ahead. With rats, if a sizeable control group is used it can provide a reasonably good estimate of what another initially well-matched group would have weighed had no manipulations occurred. Individual curves are much harder to project with confidence (see also Szepesi 1980).

Activity and Anorexia: A Possible Vicious Circle

Weight loss and activity, once initiated, could reciprocally strengthen each other, as Epling et al. (to be published) have realized. A number of species become more active when deprived of food (e.g., Cornish and Mrosovsky 1965; Collier 1969). Deprivation-induced activity could then depress eating further, and in theory a vicious circle could arise. There are at least three ways in which activity could increase weight loss. First there is the increased energy expenditure; this is reflected in the common practice of bed rest for anorectic patients. Second, activity could depress appetite and the initiation of eating. Third, it might increase satiety. The latter effect has been shown in rats allowed food for only 1 h a day and access to a wheel at other times (Kanarek and Collier 1983). The main problem facing rats subjected to a restricted feeding schedule is to override the satiety signals that arise as the hour progresses. If the hour is split into four spaced 15-min opportunities to feed, then they eat more. Wheel running at other times of day does not depress intake during four 15-min sessions but does during a 1-h session. So it appears that activity can interfere with the ability to override the satiety that develops during an hour of intense eating. These are rather special conditions. When rats are given unrestricted access to both food and exercise wheels they do eat somewhat less (e.g., Levitsky 1970), but weight losses are minor and nowhere near the threshold for starvation. Whether the hyperactivity in anorexia nervosa depresses appetite is not established. But tests could be made. Intakes on days of rest could be compared with those on days when exercise was permitted. It would not be altogether surprising if anorectics ate more rather than less after exercise — they might feel safer about not gaining weight on those days.

Conquering Homeostasis

Changes in adipocyte metabolism, CNS transmitters, body weight set points, and activity might perhaps contribute to the maintenance of the disorder, but it has to be emphasized that physiological homeostasis is by no means always dominant in the control of behavior. Even in the case of body temperature, sometimes considered as an example of a well-regulated variable, it is simple enough to strain the thermoregulatory mechanisms by offering appropriate rewards. For instance, paying men $ 0.02−0.40 for each minute they remain in a cold room, and signalling their gains prominently on a digital display, is sufficient to keep them there despite scanty dress, intense shivering, and drops in skin temperature; the higher the rate of reward, the longer they tolerate the cold (Johnson and Cabanac 1983). Turning to weight regulation, Sims and Horton (1968) induced people to become fatter than they would have done normally. Obviously the inmates of the Vermont State prison who volunteered for this experiment anticipated receiving good conduct points or some other kind of reward. With anorexia nervosa it may be more fruitful to try to understand the reward system that is powerful enough to overcome the wisdom of the body rather than to search for changes in the homeostatic machinery.

More insight may derive from studying other situations where people refuse food, even to the point of starvation, for instance in religious fasts and hunger strikes. The latter are particularly pertinent in the light of current views that feelings of ineffectiveness and lack of control are central to anorexia nervosa. Prisoners are almost totally without control over their lives. To go on hunger strike is one way they can further their causes and regain a measure of influence on those around them, even throughout the world with media amplification. Bruch's title for her recent book on anorexia nervosa, *The Golden Cage* (1978), may be more than metaphor. Whether the prison has real iron bars or is an imaginary cage of parental expectations and adolescent inadequacy, is perhaps not critical. The quickest way of course to get a person to break a hunger strike is to give in, to let them win concessions. It is a task primarily for psychiatrists, behavioral psychologists, and family therapists to find acceptable ways of making a healthy life more rewarding for the anorectic patient; and this can be very difficult when being thin itself and all that it stands for have come to be the dominant rewards. Work with animals can help define the course of anorexia nervosa once it is initiated, and some physiological complications and sustaining factors, but can play only a secondary role in combatting this uniquely human disorder.

Acknowledgements. I thank Peter Herman and Richard Keesey for commenting on the manuscript. Support came from the Natural Sciences and Engineering Council of Canada and the Atkinson Charitable Foundation.

References

Armstrong S, Coleman G, Singer G (1980) Food and water deprivation: changes in rat feeding, drinking, activity and body weight. Neurosci Biobehav Rev 4: 377–402
Bertrand HA, Masoro EJ (1977) Post-weaning food restriction reduces adipose cellularity. Nature 266: 62–63
Bjorntorp P, Yang M-U (1982) Refeeding after fasting in the rat: effects on body composition and food efficiency. Am J Clin Nutr 36: 444–449
Boyle PC, Storlien LH, Keesey RE (1978) Increased efficiency of food utilization following weight loss. Physiol Behav 21: 261–264
Bruch H (1978) The golden cage. Harvard University Press, Cambridge
Collier G (1969) Body weight loss as a measure of motivation in hunger and thirst. Ann NY Acad Sci USA 157: 594–609
Cornish ER, Mrosovsky N (1965) Activity during food deprivation and satiation of six species of rodent. Anim Behav 13: 242–248

Coscina DV, Dixon LM (1983) Body weight regulation in anorexia nervosa: insights from an animal model. In: Darby P, Garfinkel P, Garner D, Coscina D (eds) Anorexia nervosa: recent developments. Liss, New York, pp 207–219
Epling WF, Pierce WD, Stefan L (to be published) A theory of activity based anorexia. Int J Eating Dis
Faust IM, Johnson PR, Stern JS (1978) Diet-induced adipocyte number increase in adult rats: a new model of obesity. Am J Physiol 235: 279–286
Feighner JP, Robins E, Guze SB, Woodruff RA, Winokur G, Munoz R (1972) Diagnostic criteria for use in psychiatric research. Arch Gen Psychiatry 26: 57–63
Garfinkel PE, Garner DM (1982) Anorexia nervosa. Brunnel Mazel, New York
Garfinkel PE, Moldofsky H, Garner DM, Stancer HC, Coscina DV (1978) Body awareness in anorexia nervosa: disturbances in "body image" and "satiety". Psychosom Med 40: 487–498
Johnson KG, Cabanac M (1983) Human thermoregulatory behavior during a conflict between cold discomfort and money. Physiol Behav 30: 145–150
Kanarek RB, Collier GH (1983) Self-starvation: a problem of overriding the satiety signal? Physiol Behav 30: 307–311
Keesey RE (to be published) A hypothalamic syndrome of body-weight regulation at reduced levels. In: Understanding bulimia and anorexia nervosa. 4th Ross conference on medical research
Leibowitz SF (to be published) Hypothalamic noradrenergic system: role in control of appetite and relation to anorexia nervosa. In: Understanding bulimia and anorexia nervosa. 4th Ross conference on medical research
Levitsky DA (1970) Feeding patterns of rats in response to fasts and changes in environmental conditions. Physiol Behav 5: 291–300
Levitsky DA, Faust I, Glassman M (1976) The ingestion of food and the recovery of body weight following fasting in the naive rat. Physiol Behav 17: 575–580
Levitsky DA, Strupp BJ, Lupoli J (1981) Tolerance to anorectic drugs: pharmacological or artifactual. Pharmacol Biochem Behav 14: 661–667
McHugh PR, Moran TH (1977) An examination of the concept of satiety in hypothalamic hyperphagia. In: Vigersky RA (ed) Anorexia nervosa. Raven, New York, pp 67–73
Miller WH, Faust IM, Goldberger AC, Hirsch J (to be published) Effects of severe long-term food deprivation and refeeding on adipose tissue cells in the rat. Am J Physiol
Mitchell JE, Bantle JP (1983) Metabolic and endocrine investigations in women of normal weight with the bulimia syndrome. Biol Psychiatry 18: 355–365
Mrosovsky N (1983) Animal anorexias, starvation, and anorexia nervosa: are animal models of anorexia nervosa possible? In: Darby P, Garfinkel P, Garner D, Coscina D (eds) Anorexia nervosa: recent developments. Liss, New York, pp 199–205
Mrosovsky N (to be published) Animal models of anorexia nervosa. In: Understanding bulimia and anorexia nervosa. 4th Ross conference on medical research
Mrosovsky N, Powley TL (1977) Set points for body weight and fat. Behav Biol 20: 205–223
Mrosovsky N, Sherry DF (1980) Animal anorexias. Science 207: 837–842
Ozelci A, Romsos DR, Leveille GA (1978) Influence of initial food restriction on subsequent body weight gain and body fat accumulation in rats. J Nutr 108: 1724–1732
Powley TL, Keesey RE (1970) Relationship of body weight to the lateral hypothalamic feeding syndrome. J Comp Physiol Psychol 70: 25–36
Quaade FK, Vaernet K, Larsson S (1974) Stereotaxic stimulation and electrocoagulation of the hypothalamus in obese humans. Acta Neurochir (Wien) 30: 111–117
Schachter S (1982) Recidivism and self-cure of smoking and obesity. Am Psychol 37: 436–444
Sims EAH, Horton ES (1968) Endocrine and metabolic adaptation to obesity and starvation. Am J Clin Nutr 21: 1455–1470
Stordy JB, Marks V, Kalucy RS, Crisp AH (1977) Weight gain, thermic effect of glucose and resting metabolic rate during recovery from anorexia nervosa. Am J Clin Nutr 30: 138–146
Stricker EM, Anderson AE (1980) The lateral hypothalamic syndrome: comparison with the syndrome of anorexia nervosa. Life Sci 26: 1927–1934
Szepesi B (1973) "Metabolic memory": effect of antecedent dietary manipulations on subsequent diet-induced responses of rats. I. Effects on body weight, food intakes, glucose-6-phosphate dehydrogenase, and malic enzyme. Can J Biochem 51: 1604–1616

Szepesi B (1980) Effect of frequency of caloric deprivation on the success of growth compensation. Nutr Rep Int 21: 479–486

Szepesi B, Epstein MG (1977) Effect of repeated food restriction-refeeding on growth rate and weight. Am J Clin Nutr 30: 1692–1702

Vigersky RA, Loriaux DL (1977) Anorexia nervosa as a model of hypothalamic dysfunction. In: Vigersky RA (ed) Anorexia nervosa. Raven, New York, pp 109–121

Walker J, Roberts SL, Halmi KA, Goldberg SC (1979) Caloric requirements for weight gain in anorexia nervosa. Am J Clin Nutr 32: 1396–1400

Weiner H (1982) Psychobiological and psychosomatic aspects of anorexia nervosa and other eating disorders. In: West LJ, Stein M (eds) Critical issues in behavioral medicine. Lippincott, Philadelphia, pp 193–215

Weller RA, Weller EB (1982) Anorexia nervosa in a patient with an infiltrating tumor of the hypothalamus. Am J Psychiatry 139: 824–825

Noradrenergic Function in the Medial Hypothalamus: Potential Relation to Anorexia Nervosa and Bulimia

S. F. Leibowitz[1]

Anorexia nervosa and bulimia are psychiatric disorders which in recent years have been associated with well-defined diagnostic criteria (Bruch 1974; Casper et al. 1980; Garfinkel and Garner 1982; Garfinkel et al. 1980; Johnson et al. 1982; Russell 1979; Strober 1981).

Both diseases are complex, heterogeneous disorders characterized by a wide range of psychological and sociological, as well as neurochemical, endocrinological, and autonomic deviations. While a considerable amount of information about their behavior, psychopathology, and pathophysiology has been accumulated in the past decade, the etiology of anorexia and bulimia, their pathogenic mechanism, and a rational treatment for them remain to be defined. Many of the endocrine and autonomic symptoms associated with anorexia have been attributed to abnormal hypothalamic function (e.g., Mecklenberg et al. 1974; Vigersky and Loriaux 1977). At least some of the hypothalamic changes may simply reflect the consequences of weight loss and caloric deprivation, or may be associated with emotional disturbances manifested by patients with anorexia or bulimia. However, the possibility still exists that specific symptoms of these disorders may actually reflect basic hypothalamic disturbances. Many patients with anorexia nervosa have a chronic condition and are repeatedly unresponsive to attempts at weight restoration. Furthermore, some of those who attain normal body weight continue to exhibit particular endocrine and neurochemical abnormalities, as well as some degree of disturbed attitude towards food and weight.

The hypothesis proposed in this present review is that certain symptoms of anorexia nervosa and bulimia may in part be manifestations of pathological sympathetic neuronal function, particularly within the medial portion of the hypothalamus. It is hypothesized that *anorexia is associated with a decrease in α-noradrenergic function* within this specific brain area, whereas *bulimia is associated with an increase in α-noradrenergic function*. In reviewing the relevant literature below, we have focused on a variety of symptoms of anorexia and bulimia and have attempted to relate these symptoms to specific neurochemical, pharmacological, and behavioral findings obtained in animals. The hazards associated with extrapolating animal data to human psychopathology cannot be overemphasized. Anorexia and bulimia are multidimensional disorders with a wide variety of interacting symptoms and complex developmental correlates. The proposed hypotheses are clearly an oversimplification of what is actually occurring and do not reflect the range of neurochemical and physiological

---

1 The Rockefeller University, 1230 York Avenue, New York, NY 10021, USA

processes that are likely to be involved. It is believed, however, that the results described below, showing striking parallels between human and animal studies, provide a heuristic framework for generating testable hypotheses regarding disturbances specifically in food-related physiological and behavioral responses. These hypotheses and the results obtained should help us to conceptualize central neurochemical deviations that may be occurring in anorexia, as well as in bulimia. They may also provide a rational foundation for utilizing specific pharmacological approaches in treatment of these disorders.

Eating Behavior: Norepinephrine in the Medial Hypothalamus

The strong urge to eat found in bulimia produces binge-eating behavior which may be characterized as a frantic, stereotyped response involving rapid consumption of large amounts of food in a short period of time (Casper et al. 1980; Garfinkel et al. 1980; Russell 1979). The bulimic episode has been associated with signs of sympathetic activation, such as increased breathing rate, sweating, basal metabolic rate, pulse rate, and tachycardia (Mawson 1974), and is believed to result more from abnormal control of satiety than from an enhancement of hunger (Russell 1979). Although similarly preoccupied with food, anorectics exhibit a decrease in daily food consumption, a reduced meal size and rate of eating which results in a lowered body weight (Bruch 1974; Garfinkel and Garner 1982; Vigersky and Loriaux 1977). A potential association between these symptoms of anorexia and central noradrenergic activity has recently been revealed by Kaye et al. (to be published). These investigators have established that anorectics who are weight-recovered but still exhibit many symptoms of anorexia nervosa (including reduced calorie intake) manifest a 50% reduction of norepinephrine levels in their cerebrospinal fluid.

Studies conducted in the rat and in other species demonstrate that central norepinephrine (NE), particularly in the medial hypothalamus, is closely linked to eating and associated behaviours. When administered directly into the hypothalamus, NE is effective in producing an eating response in fully satiated animals. When elicited at low, near-physiological doses, this response resembles normal eating behavior; at high doses, a very rapid, large, and stereotyped eating response is observed (Leibowitz 1978a, b, 1980). Anatomical, pharmacological, and physiological studies indicate that this effect occurs specifically in the medial hypothalamus, most particularly the paraventricular nucleus; that it is mediated by α-noradrenergic receptors in this brain area; and that it is accompanied by changes in heart rate, gastric acid secretion, and peripheral glucose metabolism (Leibowitz 1980). Detailed meal pattern analyses reveal that the NE-induced potentiation of total calorie intake is attributed to an increase specifically in the size and duration of an individual meal and in the rate of eating, as opposed to any change in the frequency of the meals consumed (Grinker et al. 1982). The α-adrenergic agonist clonidine also potentiates food consumption, and chronic medial hypothalamic infusion of NE or clonidine effectively enhances 24-h food consumption and body weight (Leibowitz et al.

1982). An opposite pattern of eating behavior, namely a decrease in daily food intake, a decrease in the size and duration of each meal, and a consequent reduction of body weight, has been observed with brain manipulations which specifically attenuate α-noradrenergic activation in the medial hypothalamus. Such manipulations include chronic medial hypothalamic infusion of the α-adrenergic antagonist phenoxybenzamine, and neurotoxin or electrolytic lesions which destroy noradrenergic projections innervating the medial hypothalamus (Leibowitz and Brown 1980a, b; Leibowitz et al. 1982; Rossi et al. 1982). These findings reveal that opposite patterns of eating behavior, such as those observed in anorexia and bulimia, may occur in animals with direct manipulations, either an increase or a decrease, of medial hypothalamic α-noradrenergic activity.

Carbohydrate Ingestion

A variety of studies have indicated that anorexia and bulimia are associated with particular aberrations in the regulation of carbohydrate ingestion (Casper et al. 1980; Crisp 1967; Garfinkel et al. 1980; Lacey and Crisp 1980; Mitchell et al. 1981; Russell 1979; Vigersky and Loriaux 1977). Anorectics specifically avoid carbohydrate-rich foods, whereas bulimics focus their binge eating on palatable foods that are generally high in carbohydrate. Drugs known to potentiate food intake and hunger ratings in humans have also been found to selectively potentiate preference for carbohydrate. These drugs are 2-deoxy-D-glucose (Thompson and Campbell 1977), amitriptyline (Paykel et al. 1973), and clomipramine (Lacey and Crisp 1980), which are believed to act in part through brain α-noradrenergic systems (Leibowitz 1980; Leibowitz et al. 1978a, b; and see below).

Using a dietary self-selection feeding paradigm, experiments conducted in rats have revealed that NE injected directly into the medial hypothalamus selectively enhances ingestion of carbohydrate (Tretter and Leibowitz 1980). A similar effect has been observed with central and peripheral clonidine administration (Leibowitz et al,, to be published a; Tretter and Leibowitz 1980), as well as with peripheral injection of 2-deoxy-D-glucose (Kanarek et al. 1983). 2-Deoxy-D-glucose, which inhibits intracellular metabolism of glucose, has been found to increase the release and turnover of medial hypothalamic NE (Leibowitz 1980; McCaleb and Myers 1982; McCaleb et al. 1979). This and other evidence suggests that this feeding-stimulatory compound may act partly through the hypothalamic α-noradrenergic system. Consistent with this suggestion are the various studies which demonstrate that electrolytic or neurotoxin lesions of ascending noradrenergic afferents to the medial hypothalamus significantly reduce carbohydrate ingestion and block the stimulatory effects of 2-deoxy-D-glucose and amitriptyline on food ingestion (Leibowitz and Brown 1980a, b). Whereas these data reveal an impact of medial hypothalamic noradrenergic activity on carbohydrate ingestion, there is additional information which suggests that carbohydrate ingestion may itself specifically alter medial hypothalamic NE (Myers and McCaleb 1980). Gastric load of a carbohydrate-rich nutrient inhibits endogenous NE release in the medial paraventricular nucleus, while leaving unchanged or having the opposite effect on NE release in other hypothalamic areas. This convergence of evidence reveals a potentially important role for medial hypothalamic NE in the regulation of carbohydrate ingestion.

Antidepressant Drugs

Affective disorders have long been associated with deviations in brain noradrenergic function, and recent findings emphasize a close relationship between affective disorders and anorexia (Cantwell et al. 1977; Morgan and Russell 1975; Strober 1981) and bulimia (Hudson et al. 1982; Russell 1979; Strober 1981). The similarities in symptoms observed between depressed and eating-disordered patients have led investigators to examine the effects of antidepressant drugs in anorectics or bulimics. Several studies with tricyclic antidepressants and monoamine oxidase inhibitors have revealed a possible beneficial influence on various eating behavioral symptoms of anorexia and bulimia, as well as on mood (Johnson et al., to be published; Lacey and Crisp 1980; Paykel et al. 1973; Pope and Hudson 1982; Szmukler 1982). As mentioned earlier, the impact of tricyclic antidepressants on eating behaviour in humans appears to be accompanied by a specific enhancement of carbohydrate preference (Lacey and Crisp 1980; Paykel et al. 1973).

In the rat, tricyclic antidepressants and the monoamine oxidase inhibitor tranylcypromine are similarly found to enhance food intake, most effectively when administered directly into the paraventricular nucleus of the medial hypothalamus (Leibowitz et al. 1978a, b). This response is attenuated by α-adrenergic receptor blockers and also by NE synthesis inhibitors, reflecting a dependence on the release of endogenous hypothalamic NE. In self-selection feeding conditions, the animals exhibit a marked preference for carbohydrate after administration of the antidepressant drugs (Leibowitz et al., to be published a; Tretter and Leibowitz 1980). The effectiveness of these compounds is lost in animals with discrete neurotoxin or electrolytic lesions that destroy noradrenergic projections to the medial hypothalamus (Leibowitz and Brown 1980a).

Adrenal Hormones

The available evidence indicates that, in anorexia nervosa, the hypothalamic-pituitary-adrenal system is increased in activity (Vigersky and Loriaux 1977; Walsh et al. 1978). Plasma cortisol levels and urinary-free cortisol excretion are high. The cortisol production rate appears to be elevated. The relationship of these hormonal changes to the disturbed patterns of eating behavior in anorectics has not been determined. However, in normal human subjects there is some evidence suggesting that cortisol release, at certain points in the circadian rhythm, may be linked to eating behavior (Follenius et al. 1982; Quigley and Yen 1979). Furthermore, administration of 2-deoxy-D-glucose produces an eating response, particularly of carbohydrate, which is correlated in magnitude with the secretion of cortisol (Thompson and Campbell 1977).

A similar relationship between adrenal glucocorticosteroid release and food consumption has also been detected in animals. Studies of circadian rhythms of corticosterone in food-satiated and food-restricted rats have generally shown that plasma corticosterone concentrations reach their zenith just prior to the onset of feeding (see Leibowitz 1980 for review). In addition, peripheral injections of relatively low doses of glucocorticoids may increase food intake and weight gain in both adrenalectomized and intact animals, whereas high doses may decrease food intake. Furthermore, adrenalectomy under certain conditions may suppress food consumption, specifically when the diet contains carbohydrate. There is evidence to suggest that glucocorticoids may be closely linked with medial hypothalamic noradrenergic function in the process of controlling eating behavior, particularly ingestion of carbohydrate (Leibowitz et al., to be published c). Eating induced by medial hypothalamic NE, which is associated with enhanced carbohydrate preference, is significantly potentiated by corticosterone administration. In adrenalectomized rats the NE-elicited food intake response is abolished, but it can be completely restored by peripheral administration of corticosterone (as opposed to other steroids examined). In light of the evidence that food consumption induced by 2-deoxy-D-glucose in humans is correlated with cortisol release (Thompson and Campbell 1977), it is noteworthy that the 2-deoxy-D-glucose eating response observed in rats (and associated with NE release) is abolished by lesions which destroy specific noradrenergic projections to the medial hypothalamus (Leibowitz and Brown 1980b). This convergence of evidence suggests the possibility that the established interaction between adrenocortical steroids in control of circulating glucose homeostasis may have a parallel function in the brain, where glucocorticosteroids and medial hypothalamic NE interact to regulate food ingestion, specifically carbohydrate selection.

Insulin Response

With regard to questions of carbohydrate metabolism in anorexia nervosa, there is evidence to suggest that glucose tolerance is impaired. Additional results demonstrate insulin resistance in anorectics, which appears to persist after weight restoration despite normalization of the impaired glucose tolerance (Crisp et al. 1967). Whereas insulin has long been associated with increased appetite in humans, the implications of these changes in insulin sensitivity for the eating patterns of anorectic patients has yet to be investigated.

Although insulin-induced hyperphagia in rats has received a great deal of attention over the past decade, its underlying physiological or neurochemical substrate remains obscure. There is some evidence, however, which provides a tentative link between plasma insulin and medial hypothalamic α-noradrenergic function. Insulin, which is known to increase total calorie intake in rats, has recently been shown to preferentially increase carbohydrate ingestion (Kanarek et al. 1980). This pattern of nutrient selection is similar to that demonstrated for medial hypothalamic injection of NE (see above). Insulin potentiates the release of NE in the medial hypothalamus (McCaleb et al. 1979), and lesions which destroy medial hypothalamic noradrenergic innervation abolish the eating response induced by insulin (Leibowitz and Brown 1980b). Whereas this evidence might implicate central NE as a potential mediator of insulin-induced hyperphagia, it is premature to draw such a conclusion, particularly in light of insulin's complex interaction with peripheral metabolic processes. Some relationship, however, perhaps synergistic in nature, appears to exist between insulin and medial hypothalamic NE, as demonstrated by the finding that NE-induced hyperphagia is abolished by subdiaphragmatic vagotomy and by peripheral injection of cholinergic (muscarinic) antagonists (Sawchenko et al. 1981). Selective damage to the celiac branch of the vagus, in contrast to the gastric and hepatic branches, is similarly effective in attenuating the NE eating response. This finding has particular significance in light of the discovery that medial hypothalamic injection of NE produces a rapid release of pancreatic insulin (de Jong et al. 1977).

Circadian Rhythm

In anorexia nervosa, disturbed rhythms of plasma cortisol and luteinizing hormone have been detected (Garfinkel et al. 1975; Vigersky and Loriaux 1977). Furthermore, the circadian pattern of eating behavior in anorectics appears to be altered, with binge-eating episodes generally occurring at night. Although the basis for these findings is unknown, it is of interest that in rats the turnover of brain NE (Manshardt and Wurtman 1968) and the concentration of hypothalamic α-noradrenergic receptors (Kafka et al. 1981) follow a circadian rhythm, which is characterized by enhanced NE turnover and α-adrenergic receptor activity at a time when eating behavior is normally at its maximum. For nocturnal animals such as the rat, this occurs at the beginning of the dark cycle, when carbohydrate (relative to protein) consumption is preferentially increased (Johnson et al. 1978). Although NE-induced feeding appears to be linked in some fashion to a circadian rhythm, the precise pattern of responsiveness and the underlying determinant have yet to be defined (Leibowitz 1978b).

Food Restriction

As described earlier, anorexia and bulimia are syndromes which involve self-starvation as a predominant feature. In contrast to true anorectics, who systematically restrict their food intake to the point of life-threatening emaciation, bulimics exhibit frequent episodes of binge eating (particularly of carbohydrate-rich foods), which may be alternated with periods of self-starvation. Numerous studies conducted in the rat demonstrate that dramatic changes in brain noradrenergic activity, particularly in the medial hypothalamus, occur as a consequence of food restriction. Food deprivation, which is known to release corticosterone and to cause a preferential increase in carbohydrate and fat ingestion (McArthur and Blundell 1982), has been shown to enhance the turnover of endogenous NE in the medial hypothalamus (Martin and Myers 1975; Stachowiak et al. 1978; van der Gugten et al. 1977). Furthermore, rats deprived of food for as little as 6 h exhibit a dramatic down-regulation of α-adrenergic receptors in the medial hypothalamus (particularly the paraventricular nucleus), which is perhaps consequent to increased NE turnover, and no change or an up-regulation of receptors in numerous other brain areas examined (Jhanwar-Uniyal et al. 1980, 1982). Although there is no basis for predicting whether such noradrenergic receptor changes might also occur in humans during starvation, the magnitude and anatomical specificity of these effects and the fact that only a few hours of deprivation are required to trigger them, indicate that they may have certain physiological relevance to natural adaptive processes. This suggestion is supported by the additional evidence showing that damage to noradrenergic afferents to the medial hypothalamus, which are involved in the mediation of NE- and antidepressant-induced eating, interferes with normal carbohydrate ingestion and inhibits compensatory feeding responses which normally occur consequent to food restriction (Leibowitz and Brown 1980b). The marked down-regulation of α-noradrenergic receptors in the medial hypothalamus that develops with food deprivation (see above) can be readily reversed by just a few hours of refeeding (Jhanwar-Uniyal et al. 1982).

Water Homeostasis

Several studies have demonstrated that anorexia nervosa is accompanied by alterations in water balance. The most predominant feature appears to be an increase in urine output, associated with changes in water ingestion (Dally 1969; Mecklenberg et al. 1974; Vigersky and Loriaux 1977; Vigersky et al. 1976). A recent study of CSF and plasma vasopressin levels in anorectics has shown an increase in CSF vasopressin and CSF/plasma vasopressin ratio, in addition to a disturbed pattern of vasopressin release in response to hypertonic saline (Gold et al. 1983).

In the rat and in other species, brain NE appears to play an important role in control of vasopressin secretion. Although evidence with peripheral NE administration indicates an inhibitory function, most studies with central NE, particularly in the medial hypothalamic paraventricular nucleus and the supraoptic nucleus, suggest that this amine acts to stimulate vasopressin release (Leibowitz 1980). Norepinephrine injected into the paraventricular or supraoptic nuclei, at remarkably low doses, causes hydrated rats to concentrate their urine. Centrally injected α-adrenergic receptor antagonists inhibit this response, and at higher doses produce the opposite effect, namely an increase in urine excretion. Medial hypothalamic injection of NE also affects water ingestion, first eliciting what appears to be a natural preprandial drinking response and then inducing a profound suppression of drinking (Leibowitz 1975, 1980). Both responses are inhibited by α-adrenergic receptor antagonists. The importance of noradrenergic afferents to the medial hypothalamus in regulation of water balance is reflected by studies showing that lesions of these afferents result in enhanced water intake and altered urine excretion (Leibowitz and Brown 1980a). In light of the deficits in vasopressin release observed in anorectic patients under hyperosmotic conditions (Gold et al. 1983), it is particularly noteworthy that hypertonic saline-induced vasopressin secretion in rats appears to be mediated in part through hypothalamic α-adrenergic receptor systems (deWied and László 1967; Bridges and Thorn 1970; Leibowitz 1980).

Thermoregulation

Closely linked to disordered urine-concentrating ability in anorexia nervosa have been disturbances in temperature regulation. In anorectic patients, reduced basal temperature and abnormal thermoregulation in both hot and cold environments have been observed (Vigersky and Loriaux 1977; Vigersky et al. 1976). Extensive work in the rat and other species indicates that the defense of an organism's body temperature against cold or heat may be, in part, under the control of an α-noradrenergic system in the medial hypothalamus, specifically the anterior hypothalamus-preoptic area (Myers 1981). Norepinephrine injected into this hypothalamic area causes hypothermia, which is antagonized by α-adrenergic receptor blockers. Furthermore, body heating releases endogenous hypothalamic NE, in contrast to local neurotoxin injections, which deplete NE and impair thermoregulation against heat.

Conclusion

The evidence summarized above, obtained in clinical studies of eating disorders and in animal studies of brain neurochemical processes controlling food intake, clearly converges in support of the proposed hypothesis that certain symptoms of anorexia nervosa and bulimia may be associated with specific disturbances in medial hypothalamic α-noradrenergic function. Whereas this hypothesis concerning central NE processes is based on a particularly strong foundation, there is further evidence that other neurochemical systems, such as the opiate system, are also involved in the control of food intake (e.g., Leibowitz an Hor 1982) and may be disturbed in patients with anorexia nervosa (Kaye et al. 1982). In addition to considering these other brain neurotransmitter systems with respect to their potential role in modulating eating patterns, we will ultimately need to evaluate the relation of these systems to other symptoms of anorexia nervosa, psychological as well as endocrinological. When additional evidence becomes available it may be possible to extend the above hypothesis of noradrenergic activity to explain a variety of hormonal changes that occur in anorexia nervosa, namely an increase in cortisol and a decrease in thyroxine and gonadotrophin. The hypothesis may also help to explain changes in mood state (e.g., depression and increased activity) which characterize anorexia nervosa and have been linked to changes in central noradrenergic activation. For the moment, however, it appears most fruitful to focus simply on the changes in eating behavior, which may not only be symptoms of psychiatric disturbance but may also have profound repercussions on the overall progress of the syndrome. In this light, focusing on the bulimic episode may be particularly helpful since it involves a discrete series of events (generally lasting 2–3 h) and is therefore subject to more systematic experimental manipulation and short-term study. Furthermore, the bulimic episode may be more susceptible to manipulation since, to the anorectic patient, it appears to represent the most troublesome feature of the syndrome, in contrast to the anorexia, which is denied as being a problem and is actually welcomed and actively enforced.

References

Bridges TE, Thorn NA (1970) The effect of autonomic blocking agents on vasopressin release in vivo induced by osmoreceptor stimulation. J Endocrinol 48: 265–276

Bruch H (1974) Eating disorders: obesity, anorexia nervosa and the person within. Routledge and Kegan, London

Cantwell DP, Sturzenberger KS, Burroughs J, Salkin B, Green JK (1977) Anorexia nervosa, an affective disorder. Arch Gen Psychiatry 34: 1087–1093

Casper RC, Eckert ED, Halmi KA, Goldberg SC, Davis JM (1980) Bulimia: its incidence and clinical importance in patients with anorexia nervosa. Arch Gen Psychiatry 37: 1030–1035

Crisp AH (1967) The possible significance of some behavioral correlates of weight and carbohydrate intake. J Psychosom Res 11: 117–131

Crisp AH, Ellis S, Lowy C (1967) Insulin response to a rapid intravenous injection of dextrose in patients with anorexia nervosa and obesity. Postgrad Med 43: 97–102

Dally PJ (1969) Anorexia nervosa. Heinemann, London

deJong A, Strubbe JH, Steffens AB (1977) Hypothalamic influence on insulin and glucagon release in the rat. Am J Physiol 233: E380–E388

deWied D, László FA (1967) Effect of autonomic blocking agents on ADH-release induced by hyperosmoticity. J Endocrinol 38: 16

Follenius M, Brandenberger G, Hietter B (1982) Diurnal cortisol peaks and their relationships to meals. J Clin Endocrinol Metab 55: 757–761

Garfinkel PE (1974) Perception of hunger and satiety in anorexia nervosa. Psychol Med 4: 309–315

Garfinkel PE, Garner DM (1982) Anorexia nervosa: a multidimensional perspective. Brunner/Mazel, New York

Garfinkel PE, Brown GM, Moldofsky H, Stancer HC (1975) Hypothalamic-pituitary function in anorexia nervosa. Arch Gen Psychiatry 32: 739–744

Garfinkel PE, Moldofsky H, Garner DM (1980) The heterogeneity of anorexia nervosa: bulimia as a distinct subgroup. Arch Gen Psychiatry 37: 1036–1040

Gold PW, Kaye W, Robertson GL, Ebert M (1983) Abnormalities in plasma and cerebrospinal fluid arginine vasopressin in patients with anorexia nervosa. N Engl J Med 308: 1117–1123

Grinker J, Marinescu C, Leibowitz SF (1982) Effects of central injections of neurotransmitters and drugs on freely-feeding rats. Neurosci Abstr 8: 604

Hudson JI, Laffer PS, Pope HG Jr (1982) Bulimia related to affective disorder by family and response to the dexamethasone suppression test. Am J Psychiatry 139: 685–687

Jhanwar-Uniyal M, Dvorkin B, Makman MH, Leibowitz SF (1980) Distribution of catecholamine receptor binding sites in discrete hypothalamic regions and the influence of food deprivation on binding. Neurosci Abstr 6:2

Jhanwar-Uniyal M, Fleischer F, Levin BF, Leibowitz SF (1982) Impact of food deprivation on hypothalamic α-adrenergic receptor activity and norepinephrine (NE) turnover in rat brain. Neurosci Abstr 8:711

Johnson CL, Stuckey MK, Lewis LD, Schwartz DM (1982) Bulimia: a descriptive survey of 316 cases. Int J Eating Dis 2:3–16

Johnson CL, Stuckey MK, Mitchell J (to be published) Psychopharmacological treatment of anorexia nervosa and bulimia: review and synthesis. J Nerv Ment Dis

Johnson DJ, Li ETS, Coscina DV, Anderson GH (1978) Different diurnal rhythms of protein and non-protein energy intake by rats. Physiol Behav 21:777–780

Kafka MS, Wirz-Justice A, Naber D (1981) Circadian and seasonal rhythms in α- and β-adrenergic receptors in the rat brain. Brain Res 207:409–419

Kanarek RB, Marks-Kaufman R, Lipeles BJ (1980) Increased carbohydrate intake as a function of insulin administration in rats. Physiol Behav 25:779–782

Kanarek RB, Marks-Kaufman R, Ruthazer R, Gualtieri L (1983) Increased carbohydrate consumption by rats as a function of 2-deoxy-D-glucose administration. Pharmacol Biochem Behav 18:47–50

Kaye WH, Pickar D, Naber D, Ebert MH (1982) Cerebrospinal fluid opioid activity in anorexia nervosa. Am J Psychiatry 139:643–645

Kaye WH, Ebert MH, Raleigh M, Lake CR (to be published) Abnormalities in central nervous system monoamine metabolism in anorexia nervosa: a preliminary report. Arch Gen Psychiatry

Lacey JH, Crisp AH (1980) Hunger, food intake and weight: the impact of clomipramine on a refeeding anorexia nervosa population. Postgrad Med J 56:79–85

Leibowitz SF (1975) Pattern of drinking and feeding produced by hypothalamic norepinephrine injection in the satiated rat. Physiol Behav 14:731–742

Leibowitz SF (1978a) Paraventricular nucleus: a primary site mediating adrenergic stimulation of feeding and drinking. Pharmacol Biochem Behav 8:163–175

Leibowitz SF (1978b) Adrenergic stimulation of the paraventricular nucleus and its effects on ingestive behavior as a function of drug dose and time of injection in the light-dark cycle. Brain Res Bull 3:357–363

Leibowitz SF (1980) Neurochemical systems of the hypothalamus: Control of feeding and drinking behavior and water-electrolyte excretion. In: Morgane PJ, Panksepp J (eds) Handbook of the hypothalamus, vol 3a. Dekker, New York, pp 299–437

Leibowitz SF, Brown LL (1980a) Histochemical and pharmacological analysis of noradrenergic projections to the paraventricular hypothalamus in relation to feeding stimulation. Brain Res 201:289–314

Leibowitz SF, Brown LL (1980b) Analysis of behavioral deficits produced by lesions in the dorsal and ventral midbrain tegmentum. Physiol Behav 25:829–843

Leibowitz SF, Hor L (1982) Endorphinergic and α-noradrenergic systems in the paraventricular nucleus: effects on eating behavior. Peptides 3:421–428

Leibowitz SF, Arcomano A, Hammer NJ (1978a) Tranylcypromine: stimulation of eating through α-adrenergic neuronal system in the paraventricular nucleus. Life Sci 23:749–758

Leibowitz SF, Arcomano A, Hammer NJ (1978b) Potentiation of eating associated with tricyclic antidepressant drug activation of α-adrenergic neurons in the paraventricular hypothalamus. Prog Neuropsychopharmacol 2:349–358

Leibowitz SF, Marinescu C, Lichtenstein SS (1982) Continuous and phasic infusion of norepinephrine (NE) into the hypothalamic paraventricular nucleus (PVN) increases daily food intake and body weight in rat. Neurosci Abstr 8:711

Leibowitz SF, Brown O, Tretter JR, Kirschgessner A (to be published a) Clonidine and tricyclic antidepressants selectively stimulate carbohydrate ingestion through noradrenergic system of the paraventricular nucleus. Pharmacol Biochem Behav

Leibowitz SF, Roossin P, Rosenn M (to be published b) Chronic norepinephrine injection into the hypothalamic paraventricular nucleus produces hyperphagia and increased body weight in the rat. Pharmacol Biochem Behav

Leibowitz SF, Roland CR, Hor L, Squillari V (to be published c) Noradrenergic feeding elicited via the paraventricular nucleus is dependent upon circulating corticosterone. Physiol Behav

Manshardt J, Wurtman RJ (1968) Daily rhythm in the noradrenaline content of rat hypothalamus. Nature 217: 574–575

Martin GE, Myers RD (1975) Evoked release of [^{14}C]norepinephrine from the rat hypothalamus during feeding. Am J Physiol 229: 1547–1555

Mawson AR (1974) Anorexia nervosa and the regulation of intake: a review. Psychol Med 4: 289–308

McArthur RA, Blundell JE (1982) Effects of age and feeding regimen on protein and carbohydrate self-selection. Appetite 3: 153–162

McCaleb ML, Myers RD (1982) 2-Deoxy-D-glucose and insulin modify release of norepinephrine from rat hypothalamus. Am J Physiol 242: R596–R601

McCaleb ML, Myers RD, Singer G, Willis G (1979) Hypothalamic norepinephrine in the rat during feeding and push-pull perfusion with glucose, 2-DG or insulin. Am J Physiol 236: R312–R321

Mecklenberg RJ, Loriaux DL, Thompson RH, Anderson AE, Lipsett MB (1974) Hypothalamic dysfunction in patients with anorexia nervosa. Medicine 53: 47–159

Mitchell JE, Pyle RL, Eckert ED (1981) Binge eating behavior in patients with bulimia. Am J Psychiatry 138: 835–836

Morgan H, Russell G (1975) Value of family background and clinical features as predictors of long-term outcome in anorexia nervosa: four year follow-up study of 41 patients. Psychol Med 5: 355–371

Myers RD (1981) Neurochemistry of thermoregulation. Psychopharmacol Bull 17: 62–66

Myers RD, McCaleb ML (1980) Feeding: satiety signal from intestine triggers brain's noradrenergic mechanism. Science 209: 1035–1037

Paykel ES, Mueller PS, De La Vergne PM (1973) Amitriptyline, weight gain and carbohydrate craving: a side effect. Br J Psychiatry 123: 501–507

Pope HG, Hudson JI (1982) Treatment of bulimia with antidepressants. Psychopharmacology 78: 176–179

Quigley ME, Yen SSC (1979) A mid-day surge in cortisol levels. J Clin Endocrinol Metab 49: 945–947

Rossi J III, Zolovick AJ, Davie RF, Panksepp J (1982) The role of norepinephrine in feeding behavior. Neurosci Biobehav Rev 6: 195–204

Russell G (1979) Bulimia nervosa: an ominous variant of anorexia nervosa. Psychol Med 9: 429–448

Sawchenko PE, Gold RM, Leibowitz SF (1981) Evidence for vagal involvement in the eating elicited by adrenergic stimulation of the paraventricular nucleus. Brain Res 225: 249–269

Stachowiak M, Bialowas J, Jurkowski M (1978) Catecholamines in some hypothalamic and telencephalic nuclei of food-deprived rats. Acta Neurobiol Exp 38: 157–165

Strober M (1981) The significance of bulimia in juvenile anorexia nervosa: an exploration of possible etiological factors. Int J Eating Disorders 1: 28–43

Szmukler GI (1982) Drug treatment of anorexic states. In: Silverstone T (ed) Drugs and appetite. Academic Press, New York, pp 159–181

Thompson DA, Campbell RG (1977) Hunger in humans induced by 2-deoxy-D-glucose: glucoprivic control of taste preference and food intake. Science 198: 1065–1068

Tretter JR, Leibowitz SF (1980) Specific increase in carbohydrate consumption after norepinephrine (NE) injection into the paraventricular nucleus (PVN). Neurosci Abstr 6: 532

van der Gugten J, de Kloet ER, Versteeg DHG, Slangen JL (1977) Regional hypothalamic catecholamine metabolism and food intake regulation in the rat. Brain Res 135: 325–336

Vigersky RA, Loriaux DL (1977) Anorexia nervosa as a model of hypothalamic dysfunction. In: Vigersky RA (ed) Anorexia nervosa. Raven, New York, pp 109–123

Vigersky RA, Loriaux DL, Anderson AE, Lipsett ME (1976) Anorexia nervosa: behavioral and hypothalamic aspects. Clin Endocrinol Metab 5: 517–535

Walsh BT, Katz JL, Levin J, Kream J, Fukushima DK, Hellman LD, Weiner H, Zumoff B (1978) Adrenal activity in anorexia nervosa. Psychosom Med 40: 499–506

Effect of Starvation on Central Neurotransmitter Systems and on Endocrine Regulation

K. M. Pirke, B. Spyra, M. Warnhoff, I. Küderling,
G. Dorsch, and C. Gramsch[1]

Introduction

Amenorrhea is one of the early symptoms of anorexia nervosa. Detailed analyses of the 24-h sleep-wake pattern of gonadotropin secretion (Pirke et al. 1979) and of the luteinizing hormone (LH) response to synthetic gonadotropin-releasing hormone (GnRH) (Beumont et al. 1976) have revealed a strict weight dependence of gonadotropin secretion in anorexia nervosa. Further support for the assumption that the decrease of gonadotropin secretion may not be a primary feature of anorexia nervosa but a consequence of starvation and weight loss came from observations in dieting healthy female subjects (Vigersky et al. 1977) and from controlled starvation experiments in female volunteers (Fichter et al., this volume). Although the hypothalamic-pituitary-gonadal axis is probably influenced by starvation at various levels, the main dysfunction occurs at the hypothalamus. This conclusion can be drawn from studies by Marshall and Kelch (1979). In their patients ovulation could be induced by applying pulsatile LHRH doses over a period of weeks.

With the aim of obtaining an understanding of the hypothalamic dysfunction brought about by starvation, we have studied the rat as an animal model. This species appears to be an adequate model, since LH and gonadal hormone secretion are depressed in the starved animal (Srebnik 1970; Root and Russ 1972; Howland and Skinner 1973; Campbell et al. 1977; Pirke and Spyra 1981). We studied the male rat, since it provides an excellent model for the hypothalamic regulation of LH secretion and of the tonic feedback action of gonadal hormones on hypothalamus and pituitary gland. In contrast, the regulation of the menstrual cycle (positive feedback) is quite different in rodents and primates (Knobil et al. 1980). Due to a very powerful central signal, cycles persist for weeks in chronically starving female rats (Beumont et al., this volume). In primates the clock triggering positive feedback action appears to be in the pelvis, not in the brain (Knobil et al. 1980).

Function of the Hypothalamic-Pituitary-Gonadal Axis in the Starving Male Rat

Male Wistar rats weighing 200–220 g were used throughout the experiments. They were maintained according to an 8 h dark/16 h light schedule. Control

---

1 Max-Planck-Institut für Psychiatrie, Kraepelinstraße 2 and 10, D-8000 München 40

The Psychobiology of Anorexia Nervosa
Edited by K. M. Pirke and D. Ploog
© Springer-Verlag Berlin Heidelberg 1984

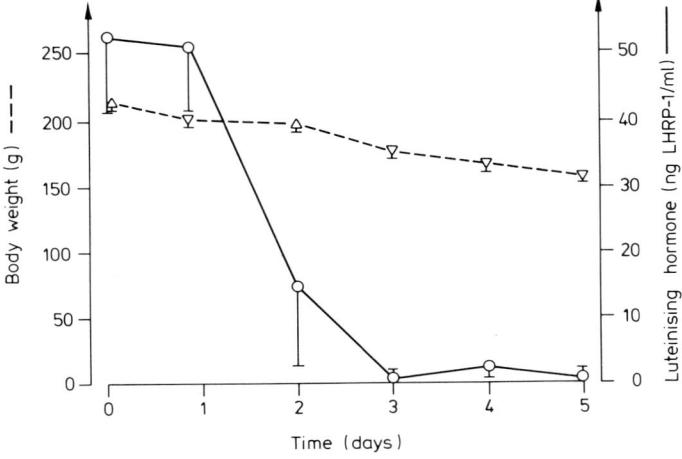

[Figure: plot of body weight (g) and luteinizing hormone (ng LHRP-1/ml) vs time (days)]

Fig. 1. Time course of body weight and plasma luteinizing hormone in starving male rats

animals were fed ad libitum with Altrumin rat food, while starved animals had access only to water.

Figure 1 shows the body weight and LH values in plasma during starvation over a 5-day period. The experimental animals lose about 25% of their body weight, while control animals gain 10−15 g during this period. The LH values are suppressed as little as 2 days after the onset of starvation and remain low thereafter. Stimulation of the Leydig cells with human chorionic gonadotropin (HCG) revealed a strong increase of plasma testosterone (Pirke and Spyra 1981), indicating an unimpaired incretory ability of the gonads. When the pituitary gland of the rat starved for 3 and 5 days was challenged with various doses of synthetic LHRH, the maximal response was unchanged after 3 days of starvation (Fig. 9). After 5 days of starvation the maximal response was decreased to about 60% compared with the control response (Pirke and Spyra 1981). The LHRH dose required to elicit one half the maximal response in LH secretion did not differ for starved and control animals (34 ng/kg body weight). All these observations clearly indicate that the hypothalamus is the level at which starvation affects the function of the hypothalamic-pituitary-gonadal axis. They further support the validity of our animal model, since starvation affects reproductive function in rat and man at apparently the same level.

The nature of the hypothalamic dysfunction responsible for the decreased LH secretion was studied. The first question addressed was whether enough LHRH is produced in the preoptic area of the starved rat. We measured the tissue content of this region, where all the perikarya of LHRH neurons are located in this species (Kawano and Daikoku 1981). In addition, we analyzed the medial basal hypothalamus and the median eminence from where the releasing hormone is secreted into the portal blood. In the preoptic area and in the basal medial hypothalamus the LHRH content was not altered by starvation (Pirke and Spyra 1981).

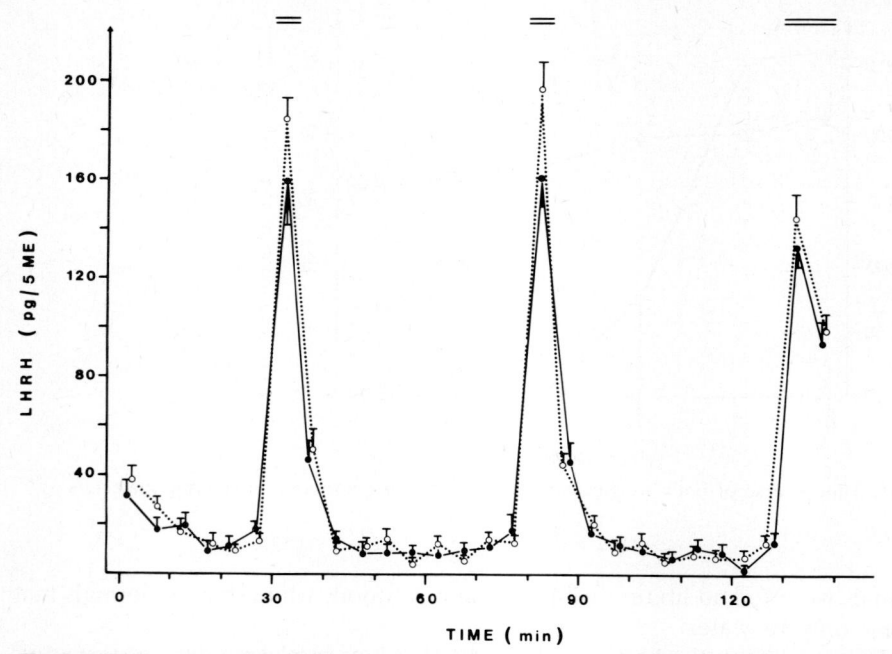

Fig. 2. LHRH release from perfused median eminences of starved *(dotted line)* and control *(solid line)* rats. The tissue was stimulated by increasing the potassium in the medium to 80 mM

The median eminence contains significantly more LHRH in the starved than in the control rat (33.2 ± 11.7 versus 15.5 ± 7.1 ng/mg protein). Although these data do not allow an estimate of the production of LHRH during starvation, they indicate that there is enough LHRH in the area from where it is normally released.

It is well known that not all the LHRH in the median eminence is immediately releasable. Only about 1% is released when the median eminence is challenged by a depolarizing stimulus in vitro (Drouva et al. 1981; Warnhoff et al. 1983). We developed an in vitro perfusion system (Warnhoff et al. 1983) in which median eminences were perfused in a chamber (volume 130 µl) at a flow rate of 100 µl/min. The tissue was challenged by depolarizing stimuli (potassium 80 mM or veratridine 50 µM).

Figure 2 shows that the LHRH release from the median eminences of starved rats is greater than in control tissue from fed rats.

Summarizing all the above-described operations, we can conclude that the impairment of the hypothalamic pituitary gonadal system in the starved rat is caused by an impaired release of LHRH from the median eminence in vivo. Since the LHRH release is closely controlled by a feedback system in which gonadal hormones regulate the gonadotropin secretion at both the hypothalamic and the pituitary level, we studied the testosterone-LH feedback in starved and control rats (Pirke and Spyra 1981). Animals were castrated and silastic capsules of different sizes filled with crystalline testosterone were implanted. These capsules release testosterone at a constant rate. A steady plasma testosterone level is achieved, which rises with increasing capsule length. Starvation was started at the same time as the animals were castrated and the capsules implanted. After 5 days animals were decapitated and testosterone and LH were measured in plasma. Starved and control animals in which substitution with testosterone was not carried out both showed greatly elevated LH values, indicating that the feedback operates during starvation. Both groups had elevated LH levels when testosterone concentrations were lower than 1.0 ng/ml. However, in the testosterone range between 1.0 and 1.8 ng/ml, 20 of 20 starved rats had very low LH values (< 10 ng/ml), while 22 of 24 fed controls had elevated LH levels (> 90 ng/ml). At testosterone levels higher than 1.8 ng/ml starved and control rats had low LH levels (< 30 ng/ml). These data indicate that the testosterone-LH feedback functions in the starved rat. The sensitivity, however, is greatly increased.

Influence of Neurotransmitters and Neuromodulators on LHRH Release from the Median Eminence

The activity of the LHRH neuron is modulated by various neurotransmitters and neuromodulators (for review see Kordan et al. 1979). Most research has been carried out on the role of noradrenergic neurons, which are assumed to affect the LHRH neuron via activation of prostaglandin E_2 (PGE_2) synthesis (Ojeda et al. 1979). The endorphinergic system is inhibitory, probably acting by inhibiting the noradrenergic neurons (Kalra 1981). This hypothesis is mainly based on the finding that the endorphin antagonist naloxone can increase LH secretion. The effect of naloxone can be inhibited by noradrenaline antagonists. Beside these three systems, others such as dopamine, serotonin, acetylcholine and various peptide modulators, e.g., vasoactive intestinal peptide (VIP), have been implicated in the regulation of LHRH production and release. Only the first three systems mentioned: noradrenaline, prostaglandin E_2, and endorphins, will be reported on below.

Noradrenaline

Noradrenaline turnover was studied in various areas of the brain, using different techniques. In the first experiments the turnover was assessed by injecting 250 mg α-methylparatyrosine (α-MT)/kg body weight IP into control rats and animals starved for 5 days. Additional starved and control rats received saline IP. α-MT blocks noradrenaline synthesis by inhibiting the tyrosine hydroxylase.

Figure 3 shows the noradrenaline concentrations in the fed and starved animals in the preoptic area and in the medial basal hypothalamus. The difference between animals receiving saline and those receiving α-MT and

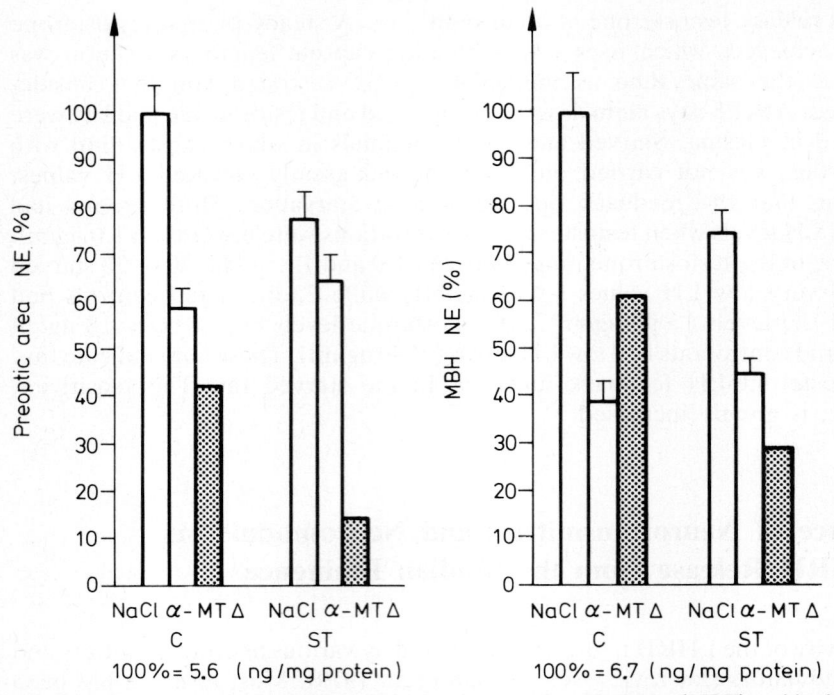

Fig. 3. Noradrenaline (*NE*) turnover in the medial basal hypothalamus (*MBH*) and in the preoptic area of starved (*ST*) and control (*C*) rats. Animals received injections of either saline or 250 mg/kg α-methylparatyrosine (α-MT) and were decapitated after 3 h. The *dotted column* presents the difference between saline- and α-MT-treated animals, which is an indicator of the turnover

sacrificed after 3 h is an indicator for the noradrenaline turnover (Weiner 1974). This technique reveals a significant reduction of noradrenaline and dopamine turnover in various areas of the brain of starved rats (Pirke and Spyra 1982).

Figure 4 shows the time course of noradrenaline turnover during starvation. After only 2 days, when the LH values in plasma are significantly decreased for the first time, the turnover is reduced and it remains constant thereafter. It is interesting to note that the noradrenaline concentrations in the preoptic area, the basal hypothalamus, and the median eminences of the saline-treated animals were only slightly reduced. This finding is complemented by the observation that there is no difference in noradrenaline secretion from the median eminences of starved and control animals when the tissue is challenged by a depolarizing stimulus (80 mM K$^+$ for 5 min) in the perfusion system described above (Fig. 5). These data indicate that the tissue concentration of noradrenaline and the releasable fraction are not decreased in starvation. The turnover, however, is significantly depressed.

Since there are some inherent problems in all methods of noradrenaline turnover measurement (Weiner 1974), we used an independent method to

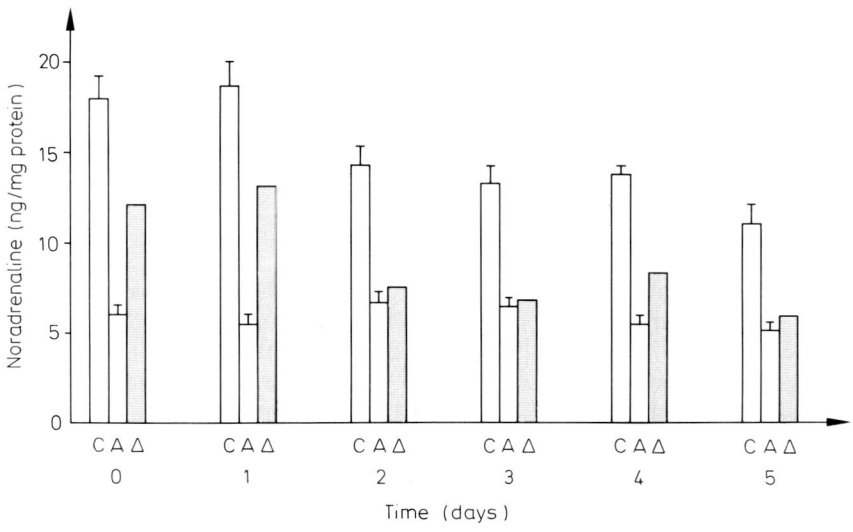

Fig. 4. Noradrenaline turnover during starvation in the male rat. The *left column* in each group (*C*) represents the saline-treated animals, the *middle column (A)* the α-methylparatyrosine-treated rats, and the *right (dark) column*, the difference between the two groups. For further explanations see text

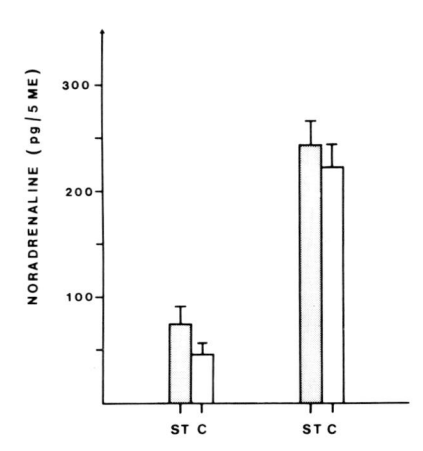

Fig. 5. Noradrenaline secretion from the median eminence of starved *(ST, dark columns)* and control *(C, light columns)* animals. The columns on the *left* of each *pair* represent basic secretion in the perfusion system. The columns on the *right* represent the noradrenaline secretion after stimulation with 80 mM potassium

Table 1. Influence of starvation on the 3-methoxy-4-hydroxyphenylglycol sulfate content[a] in the medial basal hypothalamus of starved (3 and 5 days) and control rats (pMol per medial basal hypothalamus)

	x̄	SD	n	P[a]
Control	24.3	4.6	12	–
Starved 3 days	19.3	1.9	12	< 0.01
Starved 5 days	18.9	3.0	12	< 0.01

[a] Statistical comparison was made between starved and control rats only

estimate noradrenaline turnover (Warnhoff et al., to be published). The concentration of hydroxymethoxyphenylglycol sulfate (a major metabolite of noradrenaline) was measured in the medial basal hypothalamus by high-performance liquid chromatography. The results are listed in Table 1 and indicate a significant reduction of the noradrenaline turnover after 3 and 5 days of starvation.

Postsynaptic Noradrenaline Receptors

The postsynaptic noradrenaline receptors were studied in vitro by preparing membrane preparations from hypothalamic tissues of starved (for 5 days) and control rats. For details of the technique see Spyra and Pirke (1982). Replicate experiments were performed and analyzed by the Scatchard plot technique. The data are listed in Table 2.

The receptor density was unchanged. The binding affinity, however, was significantly reduced in the starved animals in all experiments performed. Whether this increase in the dissociation constant may be responsible for the inability of noradrenergic substances to reverse the starvation-induced LH decrease (see below) remains unclear.

Pharmacological Studies with α-Adrenergic Substances

If we assume that the reduced turnover of noradrenergic neurons in the preoptic area and in the basal hypothalamus, and especially in the median eminence, is responsible for the starvation-induced LH decrease, then we should be able to stimulate LH secretion by noradrenaline agonists. This experiment, however, was unsuccessful. The suppressed LH levels in animals starved for 3 and 5 days could not be stimulated by the injection (SC) of 150 or of 300 µg clonidine/kg body weight. Neither a single injection of L-dopa (250 mg/kg) nor daily injections of 250 mg/kg for 5 days could stimulate LH secretion in the starved rat. Administration of L-DOPS was equally unsuccessful. In contrast to this negative

Table 2. Postsynaptic α-adrenoceptors in starved (5 days) and control rats

Assay[a] no.	Density of α-adrenoceptors (fmol/mg protein)		Affinity of α-adrenoceptors Kd (nM)[b]	
	Control	Starved	Control	Starved
1	116.2	108.4	0.096	0.132
2	119.8	127.2	0.107	0.144
3	100.0	104.0	0.080	0.132
Mean	112.0	113.2	0.094	0.136

[a] Assays were performed in triplicate on pooled samples (medial basal hypothalamus)
[b] Affinities (Kd) of the control and starved rats were significantly different ($P < 0.001$). The correlation coefficients of the linear curve fitting of Scatchard plots ranged from $r = 0.96$ to $r = 0.98$

finding, these substances were able to counteract the starvation-induced increase of corticosterone (Pirke and Spyra 1982).

Summarizing the findings on the noradrenergic system, we found changes in transmitter turnover and receptor affinity. The question as to whether these changes have any implications for the LHRH release from the median eminence of the starved rat, however, remains unanswered.

Endorphins

Since endorphins have an inhibitory influence on LH secretion at the hypothalamic level, we studied the tissue concentration of β-endorphin and dynorphin in the preoptic area and in the basal medial hypothalamus. Both areas were punched out of the frozen brain with a cannula (inner diameter 4 mm); the height of the tissue cylinder was 2 mm for the basal hypothalamus and 4 mm for the preoptic area. Endorphin concentrations were measured by specific radioimmunoassays, as described earlier (Gramsch et al. 1979, 1982). β-Endorphin levels in the preoptic area and in the basal hypothalamus remained unchanged after 3 and 5 days of starvation. This is in agreement with earlier studies on starved rats by Przewlocki et al. (1982).

When dynorphin was measured, a highly significant increase was seen after 3 days of starvation ($P < 0.01$) in the basal hypothalamus but not in the preoptic area (Fig. 6). After 5 days of starvation the values had returned to control levels.

The same result was obtained in female rats during starvation. (Only animals in diestrus were considered.) These observations may indicate an activation of the dynorphin system during starvation. This interpretation should be considered with great caution, however, since it is based only on measurement of tissue concentrations and not on measurement of dynorphin turnover, which is not yet measurable.

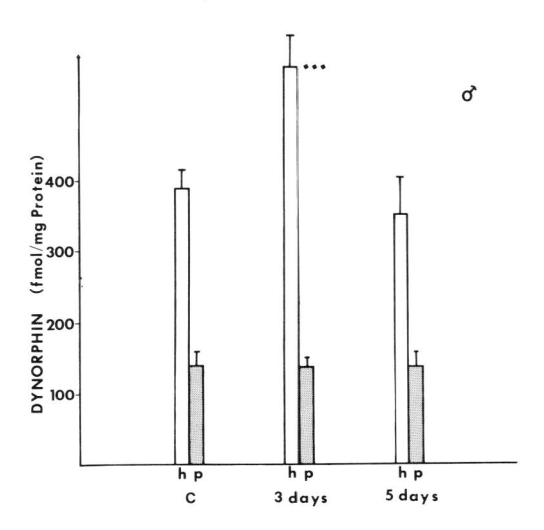

Fig. 6. Dynorphin content in control *(C)* and starved *(ST)* (3 and 5 days) rats of the medial basal hypothalamus *(h)* and the preoptic area *(p)*. After 3 days of starvation dynorphin was significantly ($P < 0.01$) increased. Twelve animals were tested in each group

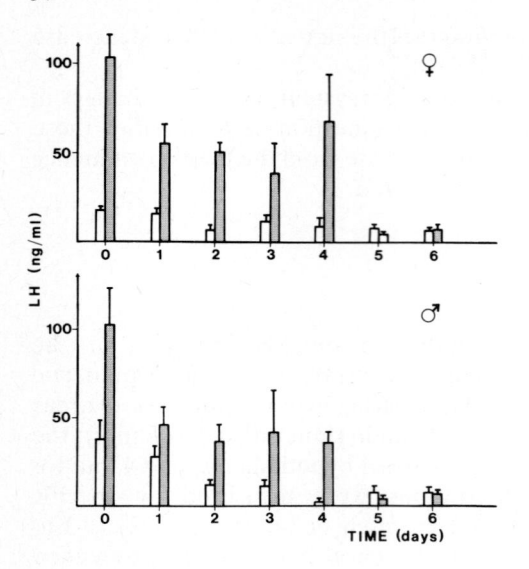

Fig. 7. Luteinizing hormone in control and starved (1–6 days) rats. The *white columns* represent saline-treated, the *dark columns* naloxone (5 mg/kg)-treated animals. Each *column* represents six animals

Another method of examining the influence of endorphinergic neurons during starvation is to antagonize the endorphinergic influence by giving naloxone (5 mg/kg body weight). We studied different groups of male and female rats (only in diestrus), which were starved for 0, 1, 2, 3, 4, 5, and 6 days. Half the group ($n = 6$) received naloxone by injection, while the other half received saline. Figure 7 shows the results.

Although starvation reduces the increase of LH after naloxone from the first day of starvation on, a stimulating effect of naloxone can be seen in animals starved for as long as 4 days. Thereafter, naloxone is no longer effective. The fact that the naloxone response in starved rats is, in all cases, smaller than in control animals does not favor the assumption that an increased activity of endorphinergic neurons is the sole cause of starvation-induced LH depression. In addition, the disappearance of the naloxone effect after 5 days of starvation indicates that other mechanisms are responsible for LH suppression. The data, however, do not exclude the possibility that at the beginning of starvation an increased dynorphin activity may contribute to the rapid suppression of LH.

Prostaglandin E_2

As already outlined, PGE_2 acts as a mediator of the stimulating influence of noradrenaline on the LHRH release (Ojeda et al. 1979). The prostaglandin synthesis depends on the supply of free fatty acids, which cannot be synthesized by the animal and have to be taken up with the food. It therefore appears reasonable to hypothesize that a shortage of PGE_2 may develop in the starving animal. It is not very informative to measure PGE_2 levels in tissue to assess the state of the PGE_2 system, since removal of the tissue sample from the brain in itself acts as a stimulus to activate PGE_2 synthesis. We have therefore studied the

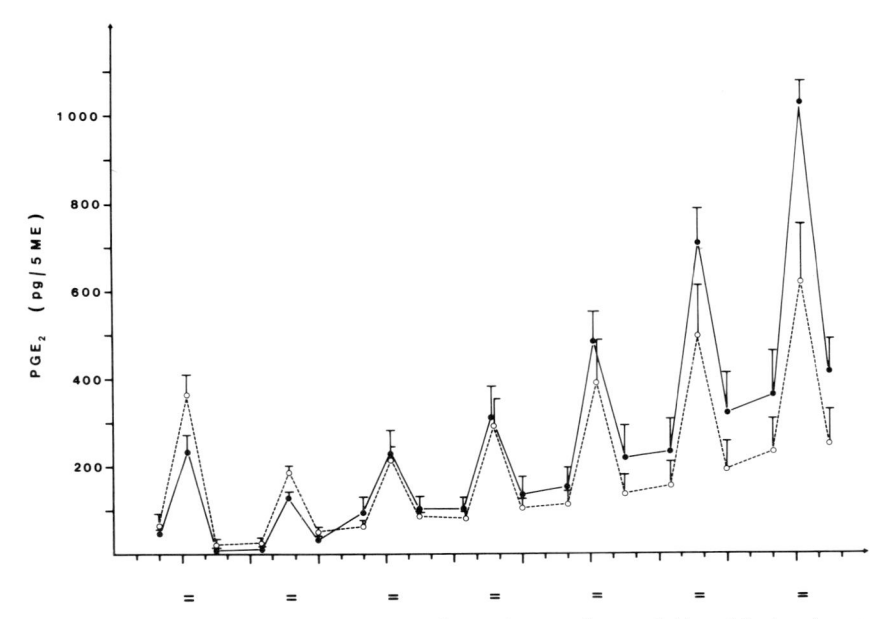

Fig. 8. Prostaglandin secretion from the median eminence of starved *(dotted line)* and control *(solid line)* rats. The tissue was challenged in an in vitro perfusion system with potassium (80 mM)

release of PGE_2 from the median eminence in the in vitro perfusion system mentioned earlier. Figure 8 shows the effect of depolarizing stimuli (80 mM K^+) on the median eminences of starved and control animals.

The first two stimuli evoke even greater responses in the starved than in the control tissue. Thereafter the basic release becomes slowly greater in starved and control samples. Peaks and baseline values become smaller in the starved tissue than in controls. These data indicate that the ability of the starved tissue to synthesize is only slightly reduced after prolonged perfusion. That means that probably enough precursor is available for PGE_2 synthesis in the starved rat. It has to be considered that the tissue from starved animals probably shows a reduced metabolism due to a general adaptation to starvation brought about, for example, by decreasing triiodothyronine and noradrenaline secretion. The slow increase of basal PGE_2 production in the perfusion experiment may be a consequence of an increasing malfunction in the perfused tissue. This may have smaller effects on tissue with a lower metabolic rate. In conclusion, the perfusion experiment does not provide clearcut evidence of an impaired function of the PGE_2 system.

In vivo LH secretion can be stimulated by PGE_2. This effect has been shown to result from an action at the hypothalamic level (Ojeda et al. 1979). We have studied the effect of PGE_2 (1 mg/kg IV) on starved and control animals.

PGE_2 can stimulate the LH release in starved rats (Fig. 9). The increase, however, becomes smaller when starvation lasts longer. This observation does not favor the assumption that a reduced ability to synthesize PGE_2 is the cause of

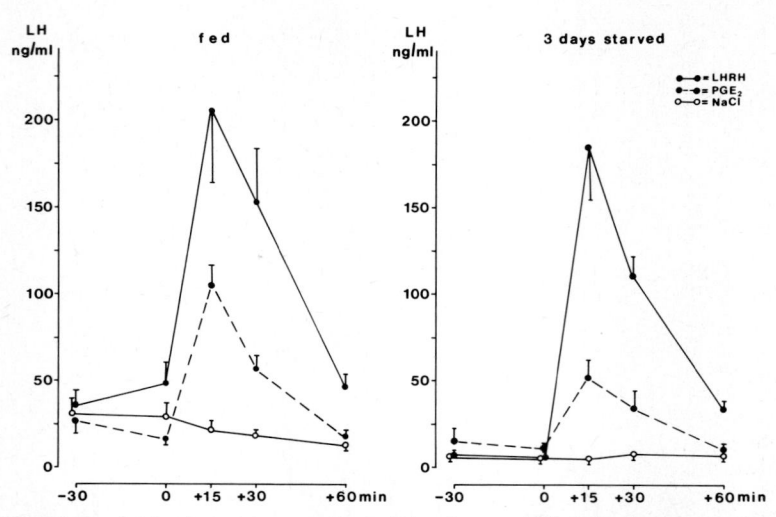

Fig. 9. Luteinizing hormone secretion in fed and starved (for 3 days) rats. Blood was drawn through an intravenous catheter. While the response to synthetic gonadotropin-releasing hormone (250 ng/kg) was unchanged, the response to PGE$_2$ (1 mg/kg) was significantly reduced ($P < 0.01$)

the starvation-induced LH decrease. There may, however, be another explanation for why the response to PGE$_2$ is decreased in starvation. We have been able to show that the effect of PGE$_2$ in the control rats depends on the state of the testosterone-LH feedback (unpublished results). Doses of testosterone which suppress LH secretion also suppress the ability of PGE$_2$ to stimulate LH release. This is not a consequence of testosterone action on the pituitary gland, since under these conditions LHRH can still evoke LH release. If we consider the altered sensitivity of the testosterone-LH feedback described earlier, the impaired response to PGE$_2$ in starved rats may well result from an altered feedback state.

In conclusion, we have described here various consequences of starvation for the noradrenergic and endorphinergic systems. These observations have not yet provided a clear understanding of the mechanisms by which starvation suppresses LHRH release from the median eminence. The information on alterations of neurotransmitters and neuromudulator function may be helpful for the interpretation of other hypothalamic dysfunctions in starvation and anorexia nervosa. As far as the effects of starvation on the LHRH release are concerned, we probably have to look into other neurotransmitter and neuromodulator systems. Another possibility is that the key to understanding will be found when the mechanism responsible for the changed sensitivity of the testosterone-LH feedback is understood.

References

Beumont PJV, George, GCW, Pimstone BL, Vinik AI (1976) Body weight and the pituitary response to hypothalamic releasing hormones in patients with anorexia nervosa. J Clin Endocrinol Metab 43: 487–496

Campbell GA, Kurcz M, Marshall S, Meites J (1977) Effects of starvation in rats on serum levels of follicle stimulating hormone, luteinizing hormone, thyrotropin, growth hormone and prolactin; response to LH-releasing hormone and thyrotropin-releasing hormone. Endocrinology 100: 580–587
Drouva SV, Epelbaum J, Hery M, Tapia-Arancibia L, Laplante E, Kordon C (1981) Ionic channels involved in the LHRH and SRIF release from rat mediobasal hypothalamus. Neuroendocrinology 32: 155–162
Gramsch C, Höllt V, Mehraein P, Pasi A, Herz A (1979) Regional distribution of methionine-enkephalin- and beta-endorphin-like immunoreactivity in human brain and pituitary. Brain Res 171: 261–270
Gramsch C, Höllt V, Pasi A, Mehraein P, Herz A (1982) Immunoreactive dynorphin in human brain and pituitary. Brain Res 233: 65–74
Howland BE, Skinner KR (1973) Effect of starvation on gonadotropin secretion in intact and castrated male rats. Can J Physiol Pharmacol 51: 759–762
Kalra AP (1981) Neural loci involved in naloxone-induced luteinizing hormone release: effects of a norepinephrine synthesis inhibitor. Endocrinology 109: 1805–1810
Kawano H, Daikoku S (1981) Immunohistochemical demonstration of LHRH neurons and their pathways in the rat hypothalamus. Neuroendocrinology 32: 179–186
Knobil E, Plant TM, Wildt L, Belchetz PE, Marshall G (1980) Control of the Rhesus monkey menstrual cycle: permissive role of hypothalamic gonadotropin releasing hormone. Science 207: 1371–1378
Kordan C, Enjalbert A, Epelbaum J, Rotsztejn W (1979) Neurotransmitter interactions with adenohypophyseal regulation. In: Gatto AM, Pech EJ, Boyd AE (eds) Brain peptides. Elsevier/North-Holland, Amsterdam, pp 277–285
Marshall JC, Kelch RP (1979) Low dose pulsatile gonadotropin releasing hormone in anorexia nervosa: a model of human pubertal development. J Clin Endocrinol Metab 49: 712–718
Ojeda SR, Negro-Vilar A, McCann SM (1979) Release of prostaglandin Es by hypothalamic tissue: evidence for their involvement in catecholamine-induced luteinizing hormone-releasing hormone release. Endocrinology 104: 617–624
Pirke KM, Spyra B (1981) Influence of starvation on testosterone-luteinizing hormone feedback in the rat. Acta Endocrinol (Kbh) 96: 413–421
Pirke KM, Spyra B (1982) Catecholamine turnover in the brain and the regulation of luteinizing hormone and corticosterone in starved male rats. Acta Endocrinol (Kbh) 100: 168–176
Pirke KM, Fichter MM, Lund R, Doerr P (1979) Twenty-four hour sleep-wake pattern of plasma LH in patients with anorexia nervosa. Acta Endocrinol (Kbh) 92: 193–204
Przewlocki R, Lason W, Konecka AM, Gramsch C, Herz A, Reid LD (1982) The opioid peptide dynorphin, circadian rhythms and starvation. Science 219: 71–73
Root AW, Russ RD (1972) Short-term effects of castration and starvation upon pituitary and serum levels of luteinizing hormone and follicle stimulating hormone in male rats. Acta Endocrinol (Kbh) 70: 665–675
Srebnik HH (1970) FSH and ICSH in pituitary and plasma of castrate protein-deficient rats. Biol Reprod 3: 96–104
Spyra B, Pirke KM (1982) Binding of (^{3}H)clonidine and (^{3}H)WB 4101 to the pre- and postsynaptic alpha-adrenoceptors of the basal hypothalamus of the starved male rat. Brain Res 245: 179–182
Vigersky RA, Andersen AE, Thompson RH, Loriaux DL (1977) Hypothalamic dysfunction in secondary amenorrhea associated with simple weight loss. N Engl J Med 297: 1141–1145
Warnhoff M, Dorsch G, Pirke KM (1983) Effect of starvation on gonadotrophin secretion and on in vitro release of LRH from the isolated median eminence of the male rat. Acta Endocrinol (Kbh) 103: 293–301
Warnhoff M, Schweiger U, Pirke KM (to be published) Noradrenaline, dopamine and serotonin turnover in the acute and chronically starved rat
Weiner N (1974) A critical assessment of methods for the determination of monoamine synthesis turnover rates in vivo. In: Usdin E (ed) Neuropsychopharmacology of monoamines and their regulatory enzymes. Raven, New York, pp 143–165

Neurotransmitter Metabolism in Anorexia Nervosa

M. H. Ebert, W. K. Kaye, and P. W. Gold[1]

Anorexia nervosa is a psychosomatic disorder in which the interplay between psychological and biological factors is particularly apparent. The illness appears to develop from a variety of psychosocial and sociocultural stressors, but when the syndrome is fully developed the symptoms are stereotyped. It is possible that, at this point, a characteristic neurobiological syndrome exists, such as occurs in endogenous depression. Evidence to support this hypothesis has developed in recent years. A variety of abnormalities of neuroendocrine function indicate hypothalamic dysfunction in the acute, underweight stages of anorexia nervosa (Vigersky and Loriaux 1977). Various neuroendocrine abnormalities documented in anorexia nervosa include abnormal regulation of growth hormone, gonadotropins, thyrotropin-stimulating hormone, cortisol, defects in urinary concentration or dilution, and failure to regulate core body temperature. Investigators in endocrinology have speculated that these endocrine abnormalities might be secondary to changes in brain neurotransmitter function or metabolism. The dexamethasone suppression test, a biological marker of endogenous depression, is abnormal in many patients with anorexia nervosa (Gerner and Gwirtsman 1981). A large amount of data in animals implicates monoamines, particularly catecholamines and serotonin, in the normal hypothalamic regulation of appetite. A catecholamine hypothesis of the etiology of anorexia nervosa and bulimia has been proposed by Leibowitz (Leibowitz, to be published).

The studies summarized in this review have been conducted in our research group over the past several years to delineate the neurobiological syndrome that exists during the acute cachexic stage of anorexia nervosa, and to determine whether there are aspects of this syndrome that persist after weight is regained. If there are persisting neurobiological changes, they might play a role in the chronic, tenacious course of the illness. In our studies we have focused on several neurotransmitter systems known to play a modulating role in appetite regulation and feeding behavior in animals. Several of these neurotransmitter systems are also implicated in hypotheses of the etiology of major affective disorders. Recently it has become apparent that a broad interface exists between major affective disorders and eating disorders. Many individuals with anorexia nervosa meet research diagnostic criteria for a major or minor affective disorder, and there is a high incidence of affective disorders in family members of patients with anorexia nervosa (Gershon et al. 1983).

---

1 National Institute of Mental Health, Department of Health and Human Services, Intramural Research Program, Bethesda, MD 2025, USA

In particular, the neurotransmitters and neuromodulators whose function and metabolism we have tried to study in patients with anorexia nervosa include norepinephrine, dopamine, serotonin, endogenous opioids, and arginine vasopressin. All these neurochemicals play a role in the hypothalamic regulation of feeding behavior in animals. Several play a key role in neuroendocrine regulation in the hypothalamus. Several of these neurochemicals counterregulate each other's function in brain; for example, norepinephrine and vasopressin systems in the hypothalamus. These substances seemed to be a logical starting point to delineate the neurobiological syndrome of anorexia nervosa.

Methods

Details on patients, treatment and on collecting blood and CSF-samples have been published elsewhere (Kaye et al. 1982; Gold et al. 1983).

Results

Catecholamine and Indoleamine Metabolism

Investigations to date of catecholamine metabolism in anorexia nervosa have been of peripheral or total-body metabolism. Plasma norepinephrine (NE) levels, both at rest and under stress, are about 50% of normal levels (Gross et al. 1979). Urinary excretion of metabolites of NE and dopamine (DA) are decreased in the underweight stage of anorexia nervosa (Gross et al. 1979). A major methodological issue in these studies has been whether or not the changes in neurotransmitter metabolism are due to starvation alone, or are related also to the psychiatric syndrome of anorexia nervosa. The present study was undertaken to investigate whether or not there was evidence for alteration in central nervous system metabolism of biogenic amines in the various stages of anorexia nervosa, including stages of the illness where body weight is not severely low.

Levels of 5HIAA and HVA in lumbar CSF were measured to estimate the turnover rate of serotonin and dopamine in the central nervous system (CNS). CSF levels of NE were measured to estimate release and turnover of NE in the CNS. Plasma amino acids were measured at the time of lumbar puncture because of evidence that precursor amino acids regulate synthesis of catecholamines and indoleamines under certain conditions.

All groups had similar blood concentrations of free tryptophan, total tryptophan, phenylalanine, and tyrosine, with the exception of a significant increase in tyrosine concentration ($t = 3.33$, $P < 0.02$) after weight recovery. The ratio of tryptophan, tyrosine, or phenylalanine to the sum of the remaining

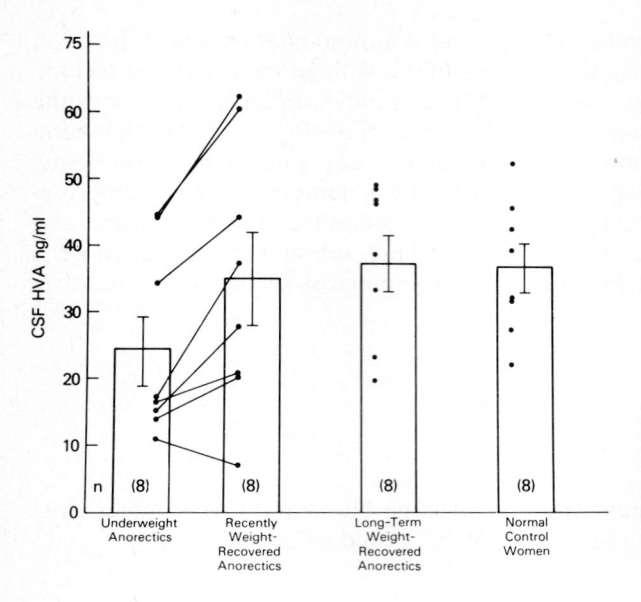

Fig. 1. Mean (± SEM) concentrations of HVA, the major CNS dopamine metabolite, in CSF for all three groups of patients with anorexia nervosa and normal controls. Units are nanograms per milliliter

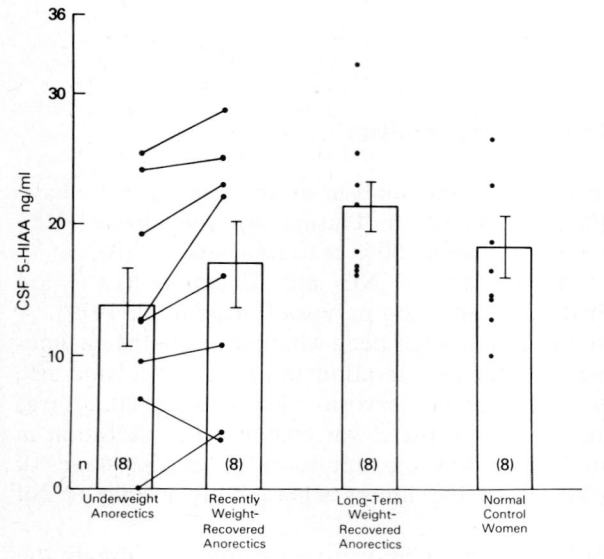

Fig. 2. Mean (± SEM) concentrations of 5-hydroxy-indoleacetic acid, the major CNS serotonin metabolite, in CSF for all three groups of patients with anorexia nervosa and normal controls. Units are nanograms per milliliter

neutral amino acids was not significantly lower for underweight anorectics than for normal controls.

Seven of eight patients studied as underweight anorectics showed an increase in CSF HVA (Fig. 1) and 5HIAA (Fig. 2) after weight recovery ($P < 0.01$ and $P < 0.05$, respectively). One patient had a decrease in CSF HVA and another in 5HIAA. Group means for CSF metabolites were otherwise similar, except that long-term weight-recovered anorectics had significantly elevated CSF 5HIAA concentrations compared with the underweight anorectics ($P < 0.02$).

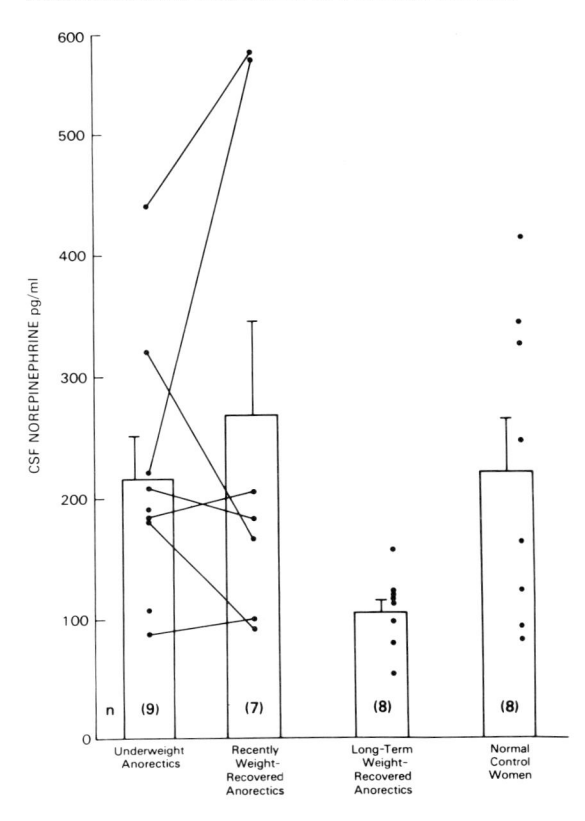

Fig. 3. Mean (± SEM) concentrations of norepinephrine in CSF for all three groups of patients with anorexia nervosa and normal controls. Units are picograms per milliliter

In contrast to the CSF metabolites, CSF NE concentrations (Fig. 3) demonstrated no consistent pattern of change between underweight anorectics and the same patients after weight recovery. Long-term weight-recovered patients had significantly lower concentrations of CSF NE than the normal controls ($P < 0.05$) and underweight anorectics ($P < 0.02$), and demonstrated a similar trend to the recently weight-recovered anorectics ($P < 0.01$).

To our knowledge, there are no previous studies of CNS catecholamine and indoleamine metabolism in anorexia nervosa. We decided to measure concentrations of CSF HVA and 5HIAA because these metabolites are the major products of dopamine and serotonin metabolism in the brain and reflect DA and 5-HT turnover in the CNS (Ebert et al. 1980). We elected to measure NE directly in CSF rather than its major metabolite, MHPG, because of evidence that a substantial portion of free MHPG in human CSF is derived from plasma (Kopin et al. 1982). Peripheral NE apparently does not cross the blood-brain barrier or the blood-CSF barrier (Perlow et al. 1978; Weil-Malherbe et al. 1959). However, it is unclear whether CSF NE concentrations adequately reflect brain NE metabolism because of rapid reuptake of NE by brain neurons (Fuxe and Ungerstedt 1966, 1968) and because lumbar CSF may disproportionately reflect spinal cord NE. A major difficulty with neurochemical studies in

CSF is that measurements of monoamines and their metabolites in CSF reflect the sum of contributions of various brain and spinal cord regions. There is no method presently available to relate changes in CSF monoamine concentrations to specific brain regions, but CSF studies remain the best technique currently available for estimating brain monoamine neurotransmitter activity.

Understanding of the relationship of these disturbances of central monoamine metabolism to the individual symptoms of anorexia nervosa is confounded by a number of problems. An initial question is whether the change in neurotransmitter metabolism is a consequence of caloric deprivation or weight loss, or is related to the psychological syndrome of anorexia nervosa.

All the underweight anorectic patients in this study were at stable weight and caloric intake for at least 2 weeks prior to study. Although total caloric intake is low in underweight anorectics, nutrient intake, when calculated as calories per kilogram of body weight per day, is within a normal range. We found that underweight anorectics had normal concentrations of monoamine precursor amino acids in their blood, compatible with the findings of Russell (1967) but in contrast to those of Coppen, who reported (Coppen et al. 1976) low total and free plasma tryptophan in underweight anorectics. It is thought that all neutral amino acids compete for one transport site at the blood-brain barrier (Wurtman 1967), and that therefore the uptake of any one neutral amino acid depends on the concentration in blood of the other neutral amino acids. For the ratios of the amino precursors of 5-HT, DA and NE, underweight anorectics had similar values to normal controls. An acute state of starvation or depletion of monoamine precursors is unlikely to account for the CSF monoamine abnormalities in our underweight anorectics.

While underweight anorectics were not in a state of starvation, chronic low weight and prior caloric deprivation might account for changes in monoamine metabolism. Studies in animals show that short-term caloric deprivation increases or leaves unchanged hypothalamic or cortical indoleamine concentrations (Curzon et al. 1972; Kantak et al. 1978). Food deprivation has been variously reported not to change (Loullis et al. 1979), to increase (Friedman et al. 1973), and to decrease NE and DA turnover (Pirke and Spyra 1982) in the rat hypothalamus and cortex. Methods of producing weight loss, measurements of brain monoamines, and areas of the brain surveyed differ among these animal studies, making it difficult to compare results. Furthermore, extrapolating from short-term animal studies to the chronic condition of anorexia nervosa is problematic.

A possibility also exists that some disturbance in monoamine function drives weight loss. Animal studies support the concept that hypothalamic monoamine systems regulate appetitive behavior. In general, DA and 5-HT systems can inhibit appetite (Leibowitz 1980). NE pathways in the hypothalamus can facilitate feeding (Ritter and Epstein 1975; Ritter et al. 1975) by stimulating hypothalamic α-adrenergic receptors (Leibowitz 1980; Booth 1967; Slangen and Miller 1969) or suppress feeding by β-2-receptor activation (Leibowitz 1980; Goldman et al. 1971). It appears that monoamine metabolism is intimately linked with control of appetite and weight, and that changes in one affect the other. An association between brain monoamine metabolism, appetitive behavior, and weight loss in anorexia nervosa appears likely, although the causal sequence of events remains unknown.

Another issue is the potential relationship between disturbances in brain monoamine metabolism and neuroendocrine disturbances in underweight anorectics. Underweight anorectics have neuroendocrine disturbances which tend to normalize with weight recovery. We found that CSF concentrations of DA and 5-HT metabolites, but not CSF NE, have a similar pattern. Could abnormalities in CNS DA and 5-HT pathways account for the neuroendocrine disturbances which tend to normalize with weight recovery? It is known that central 5-HT and DA pathways modulate some of the neuroendocrine systems that are disturbed in underweight anorectics. There is considerable pharmacological evidence that the central release of 5-HT stimulates corticotrophin-releasing hormone (CRH) secretion (Weiner and Ganong 1978), and that hypothalamic DA transmission appears to participate in the regulation of luteinizing hormone (Weiner and Ganong 1978; Smythe 1977). The evidence for an effect of central DA and 5-HT pathways on other neuroendocrine systems in less conclusive. Disturbances in central DA or 5-HT function may contribute to neuroendocrine abnormalities in underweight anorectics. However, more specific conclusions will only be possible when further studies, such as challenge with pharmacologic agents with relatively specific neurochemical actions, are done.

Of the brain monoamine systems investigated, only brain NE pathways have a role in the normal regulation of each of the neuroendocrine systems that become disturbed in underweight anorectics. NE appears to exert an excitatory influence on gonadotropin release (Weiner and Ganong 1978; Sawyer et al. 1974), perhaps by stimulating the tonic secretion of the luteinizing hormone-releasing system (Loffstrom 1977; McCann 1970). Release of NE in the hypothalamus decreases CRH secretion (Weiner and Ganong 1978) and appears to have an excitatory effect on the hypothalamic control of TSH secretion (Krulich et al. 1977; Scapagnini et al. 1977). Central NE pathways regulate hypothalamic vasopressin release (Blessing et al. 1982). Brain NE and opiate systems are intimately linked (Korf et al. 1974; Gold et al. 1979). It is known that anorectics have disturbances of mood (Cantwell et al. 1977), motor activity (Kron et al. 1978), and temperature regulation (Vigersky and Loriaux 1977). While other monoamine systems also contribute to regulation of these systems, adrenergic pathways clearly play an important role in each (Cooper et al. 1982).

In summary, decreases in CNS 5-HT and DA metabolism appear to be associated with the state of low weight. Either may contribute to some disturbances in neuroendocrine function, mood, or cognition found in underweight anorectics. CSF NE concentrations have a more complex pattern. There is no evidence in this study that concentrations of CSF NE are abnormal in the low-weight state or after recent weight recovery. However, CSF NE decreases during long-term weight recovery. These data raise the possibility that CNS NE metabolism may play a vital role in anorexia nervosa. These preliminary findings have encouraged us to further investigate monoamine function in anorexia nervosa. Studies in progress include measurements of biogenic amines in central and peripheral compartments in a larger sample of anorectics and aminergic challenges of neuroendocrine systems in anorexia nervosa.

Endogenous Opioids

Since the discovery of the opiate receptor and the endogenous opiate peptides (endorphins), there has been speculation regarding their role in human behavior. A considerable body of data, derived primarily from animal experimentation, suggests that the endogenous opioid system may be one of the neurotransmitters involved in eating behavior. β-Endorphin, when injected into the ventromedial hypothalamus, stimulated food intake in satiated rats (Grandison and Guidotti 1977), and a synthetic enkephalin, when infused into the lateral ventricle of sheep, initiates feeding in satiated animals (Baile et al. 1980). Naloxone given IP reduces food intake in food-deprived rats (Holtzman 1974) and abolishes overeating in genetically obese mice and rats (Margules et al. 1978). Pituitary tissue from obese mice and rats has been shown to contain twice as much β-endorphin as that from lean littermates.

For this study we used the measurement of total opioid activity, determined by radioreceptor assay, in CSF of patients with anorexia nervosa and of control subjects to study possible relationships between the endogenous opioid system and anorexia nervosa (Kaye et al. 1982). Since there are numerous endogenous opioids and few data that suggest differential behavioral effects or roles of any specific opioid, this strategy of measuring total opioid activity serves as an overall assessment of endogenous opioid system activity reflected by the presence of opioid in CSF. In this regard it is an analog of naloxone treatment, in which overall endogenous opioid system activity is blocked.

Levels of CSF opioid activity were determined by radioreceptor assay (Naber et al. 1980, 1981).

The mean ($\pm$ SE) level of CSF opioid activity was significantly higher in five anorectic patients at their minimum weight (6.08 $\pm$ 0.70 pmol/ml) than in 1) the same patients at their restored weight (2.04 $\pm$ 0.94 pmol/ml; $P < 0.02$); 2) eight recovered anorectic patients (1.87 $\pm$ 0.66 pmol/ml; $P < 0.01$); and 3) seven normal controls (2.98 $\pm$ 0.48 pmol/ml; $P < 0.01$) (Fig. 4). The mean level of CSF opioid activity of the eight recovered patients was lower than that of the seven normal controls, although this difference was not significant. The mean level of CSF opioid activity of the seven normal controls was comparable to that of 14 normal female controls studied previously, for whom stage of menstrual cycle was not controlled (3.18 $\pm$ 0.55 pmol/ml).

Thus significant elevations in total CSF opioid activity levels are observed in anorexia nervosa patients at their minimum weight compared with levels at their restored weight and with levels in weight-recovered subjects with a history of anorexia nervosa or from normal controls. The change in opioid activity and the improvement in self-ratings of discomfort associated with eating both accompanied weight restoration but were not significantly correlated with each other.

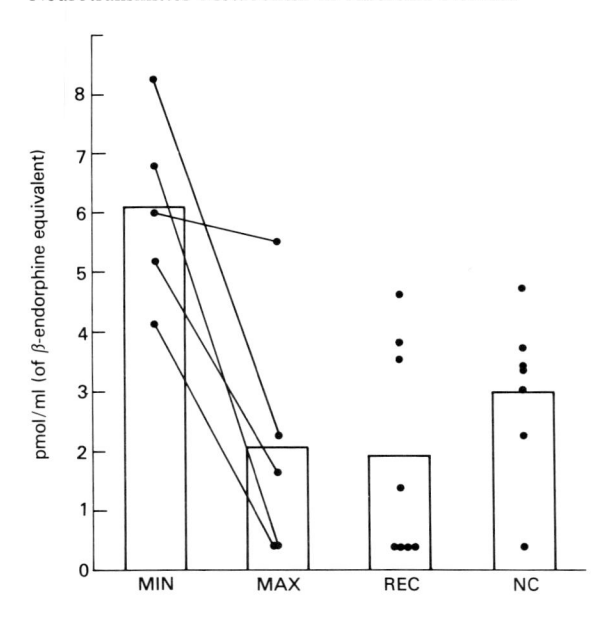

Fig. 4. Mean (± SEM) concentrations of total opioid activity in CSF for all three groups of patients with anorexia nervosa and normal controls. Units are β-endorphin equivalents per milliliter

Most data concerning endogenous opioids and appetite regulation suggest that opioid agonists stimulate eating and antagonists diminish eating. We found elevations in opioid activity only at the time of minimum weight. It is unknown whether increased opioid activity reflects a specific opioid appetite stimulation that is a consequence of diminished caloric intake or a stress effect that might serve as a protective response to decrease metabolic requirements when weight is lost.

It has been suggested that the endogenous opioid system plays an important role in the biologic response to stress. Margules (1979) has reviewed evidence that the peripheral endorphinergic system may aid survival in starvation by conservation of nutrients and water and by decreasing energy-expending activities. Reducing thyrotropin release, lowering the set point for body temperature, and decreasing respiration rate contribute to decreased metabolism in body tissues. Recently Gerner and associates (Gerner and Gwirtsman 1981) reported abnormalities in dexamethasone suppression in anorectic patients, indicating abnormal activation of the hypothalamic-pituitary-adrenal axis, a system known to be responsive to stress. Rubinow and associates (Rubinow et al., to be published) reported that CSF opioid activity was significantly correlated with urinary free cortisol excretion in depressed patients, who as a group had abnormally high cortisol excretion, but not in normal controls, suggesting that endogenous opioids may be related to abnormalities in the hypothalamic-pituitary-adrenal axis.

Central and Peripheral Vasopressin Function

The suggestion of deficient secretion of AVP in patients with anorexia nervosa is of particular interest in the context of recent evidence that AVP is widely distributed in brain beyond the boundaries of the hypothalamus (Zimmerman and Robinson 1976; Buijs et al. 1978; Sofroniew 1980) and has been reported to influence complex behavioral and cognitive functions in experimental animals and humans (de Wied 1976; Gold et al. 1978; Weingartner et al. 1981).

In our studies we have explored possible abnormalities of AVP function in patients with anorexia nervosa by examining the secretion of AVP before and at intervals after correction of the weight loss (Gold et al. 1983). Since subnormal release of AVP in the setting of dehydration has been suggested by earlier work (Vigersky et al. 1976; Mecklenberg et al. 1974), we first studied the pattern of plasma AVP secretion to an osmotic stimulus. We also measured the levels of AVP in the CSF, since there is indirect evidence of a relationship between AVP secretion into the plasma and CSF (Luerrson and Robertson 1980), and in light of circumstantial evidence which suggests that the CSF may constitute a physiologically relevant pathway which conveys hypothalamic peptides such as AVP to their disparate sites of action in brain (de Wied 1976; van Wimersma-Greidanus and de Wied 1976). Specifically, we attempted to answer three questions about AVP secretion: 1) Is the secretion of plasma AVP deficient, as previously observed using indirect methods (Vigersky et al. 1976; Mecklenberg et al. 1974), and if so, what is the nature of the defect? 2) Are there abnormalities in CSF AVP, and if so, are they related to the abnormalities in plasma AVP? and 3) Are defects in CSF and/or plasma AVP a constant feature of the syndrome or are they more likely to show a normal pattern after weight gain?

The results of these studies show that the osmoregulation of plasma AVP is abnormal in most if not all chronically underweight and acutely recovered patients with anorexia nervosa. Ths abnormality is not apparent under basal conditions but is clearly manifest when an osmotic stimulus such as a hypertonic saline infusion is administered. The abnormal response takes two forms. The least common is a simple deficiency of AVP secretion. It is characterized by a rise in plasma AVP that correlates closely with the rise in plasma sodium but is quantitatively subnormal relative to the strength of the stimulus. This pattern is found in 25% of our chronically underweight patients and is consistent with previous reports based on indirect methods suggesting that 40% of these patients have partial neurogenic diabetes insipidus. A much more common defect in our study is erratic or osmotically uncontrolled AVP release. It is characterized by a failure of any detectable increase in AVP secretion to hypertonic saline with fluctuations in plasma AVP during the infusion that bear no relationship whatsoever to changes in plasma sodium. This pattern is in marked contrast to healthy adults who invariably show a smooth, progressive rise in AVP that correlates closely with the rise in plasma sodium during the infusion of hypertonic saline. Erratic secretion of AVP does not seem to be associated with a gross deficiency of the hormone since most of the plasma AVP levels in these patients fall within the normal range. However, the two types of AVP defects

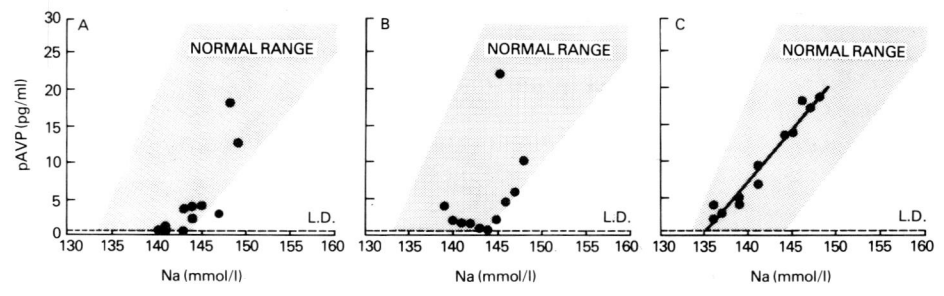

Fig. 5. Relationship between plasma AVP and plasma sodium during an IV hypertonic saline infusion in a patient with anorexia nervosa studied longitudinally while chronically underweight (**A**), acutely recovered (**B**), and chronically recovered (**C**)

may not be entirely unrelated, since one subject studied longitudinally exhibited AVP deficiency when chronically underweight and an erratic pattern when acutely recovered.

Our findings also suggest that abnormal osmoregulation of AVP is not a permanent or pathognomonic feature of anorexia nervosa but is nonetheless a strongly entrenched defect that is not easily corrected by improved nutrition. Thus, abnormal AVP secretion persisted in all the chronically underweight patients restudied soon after correction of the weight loss, but was absent in most of the patients who had maintained their weight for 6 months or more. One patient studied longitudinally in all three phases of the disorder exhibited the full transition from abnormal to normal AVP secretion (Fig. 5); a second patient studied longitudinally while chronically underweight and during chronic recovery showed a similar transition.

The pathogenesis of the plasma AVP abnormalities in our underweight and recovered anorectics remains undefined. We cannot determine whether these defects in plasma AVP secretion reflect a process unique to anorexia nervosa or are nonspecific effects of chronic inanition. The only conclusion permitted by our study is that they are not due to interference by any of the recognized or putative nonosmotic stimuli (Gold et al. 1978). During all the saline infusions, blood pressure was stable and vomiting did not occur, and the patients showed no evidence of pain, anxiety or other forms of stress. In some of the studies performed in the chronically underweight state basal blood pressure and, perhaps, blood volumes were low, but these changes would be expected to lower the set of osmostat or increase its sensitivity rather than produce erratic or deficient secretion of AVP (Robertson et al. 1976). Hence, the abnormalities observed in our patients are probably due to some intrinsic defect in either the neurohypophysis, the osmoreceptor or some of the other regulatory afferents.

Our findings also reveal that abnormalities in the osmoregulation of plasma AVP are often accompanied by abnormalities in the levels of AVP in CSF. In healthy adults or patients without evident brain disease, we find the daytime concentration of AVP in CSF ranges from 0.5 to 2.0 pg/ml. In addition, the CSF levels tend to parallel but are almost always lower than those in plasma. The latter relationship is retained in patients with SIADH (Luerrson and Robertson 1980) or depression (Gold et al. 1981) but is often reversed in patients with neurogenic diabetes insipidus (Luerrson and Robertson 1980) or neoplasms of the hypothalamic/pituitary region (Jenkins et al. 1980). In our chronically underweight patients, a bimodal distribution in CSF AVP was observed, and the four patients with shortest duration of illness each had abnormally high levels of AVP in CSF; moreover, six of the eight chronically underweight subjects had reversal of the normal CSF/plasma ratio. These findings suggest that at least half of underweight patients with anorexia nervosa have a defect in the regulation of CSF AVP, which seems to be most pronounced in subjects with a relatively recent onset of illness. The true incidence of CSF abnormalities probably is even higher because certain types of defects, such as erratic secretion, might not always be detected by measurement of a single, unstimulated basal value. A better indication of the incidence and nature of the CSF defect could be obtained were it possible to carry out more dynamic tests, such as those used to assess the secretion of plasma AVP.

The abnormalities in CSF AVP seemed to improve with a gain in weight but the change was slight and not necessarily progressive. In the four patients found to have an absolute increase in CSF AVP when chronically underweight, repeat values obtained immediately after return to ideal body weight were uniformly lower but only two of them reached the normal range. The findings during sustained recovery were no better. Of the eight patients tested, the absolute level of AVP in CSF was supranormal in three (38%), an incidence not significantly different from the acutely recovered (33%) or chronically underweight (50%) states. Moreover, the abnormal values tended to be as high as or even higher than in the other two phases of the disorder. Essentially the same pattern was observed with the CSF/plasma AVP ratios. Taken as a whole, these findings suggest that the abnormalities in CSF AVP in anorexia nervosa are less frequent but more refractory to treatment than those in plasma. However, as noted above, these slight differences in incidence and course could be due simply to the different methods used to detect abnormalities in the two compartments. The present data are more compatible with this latter concept, since the defects in CSF and plasma AVP seem closely related: thus the two defects almost always occurred together even in patients studied after a sustained weight gain.

As with the plasma AVP abnormalities, the CSF defects cannot be definitively ascribed to either a pathophysiological process intrinsic to anorexia nervosa or to possible nonspecific effects of chronic weight loss. Evaluation of the pathogenesis of the CSF defects is further complicated by the paucity of information about CSF AVP. The weight of available circumstantial evidence suggests that AVP in the CSF is not derived by diffusion from the plasma. Thus, large IV doses of AVP fail to elevate the level of AVP in the CSF (Luerrson and Robertson 1980).

Moreover, in primates, a circadian rhythm for AVP in CSF but not plasma has been detected (Reppert et al. 1981), while patients with diabetes insipidus show substantial levels of AVP in CSF but none in plasma (Luerrson and Robertson 1980). These latter studies suggest that AVP in CSF derives from brain pathways that are anatomically distinct from those which secrete into blood. This conclusion is compatible with immunohistochemical studies which suggest that extrahypothalamic AVP pathways which extend from the paraventricular nucleus to the third and lateral ventricles originate from magnocellular neurons different from those which project to the neurohypophysis (Zimmerman et al. 1976; Buijs et al. 1978).

The factors regulating the secretion of AVP in CSF are equally uncertain. However, despite the apparently different sources and temporal organization of AVP in the plasma and CSF, some stimuli seem to exert similar actions on the neurosecretory pathways which bring AVP to these two compartments. For instance, hemorrhage and excitation of the vagus increase the antidiuretic activity of both CSF and plasma (Robertson 1977). Moreover, in patients with inappropriate AVP secretion, AVP is also elevated in the two spaces, indicating that even pathological stimuli can exert a similar action on the two putative neurosecretory pathways.

In contrast to patients with diabetes insipidus, reversal of the CSF/plasma gradient in anorexia nervosa does not seem to reflect an isolated deficiency of plasma AVP: rarely did patients show low basal plasma AVP and/or deficient secretion during osmotic stimulation; rather, the prevailing pattern was that of supranormal levels of AVP in plasma. Volume contraction or diminished blood pressure also seem unlikely causes of the reversed gradient, since similar changes in experimental animals caused marked elevations of AVP in both the plasma and the CSF compartments.

The pathophysiologic consequences of the abnormal regulation of AVP in the plasma and CSF of patients with anorexia nervosa remain to be fully defined. None of the patients with documented abnormalities in osmoregulation exhibited gross abnormalities in systemic water balance, as evidenced by plasma sodium concentrations. However, most, if not all of these subjects exhibited significantly increased levels of urine output. This finding is difficult to explain in patients who show erratic secretion without obvious deficiency of vasopressin. It could reflect the cumulative effect of intermittent deficiency of plasma vasopressin or an impaired renal response to the hormone. None of the patients with polyuria showed evidence of hypokalemia, hypercalcemia, or any other recognized cause of nephrogenic diabetes insipidus.

The possible significance of the CSF abnormalities are speculative because we do not know with certainty what role, if any, AVP plays in CNS function. Since the administration of AVP has been reported to enhance the strength of trace events in memory of depressed patients (Gold et al. 1978) and normal subjects (Weingartner et al. 1981), it is conceivable that increased levels of AVP in CSF, regardless of cause, could strengthen the encoding of certain thoughts in anorectics, particularly with regard to the tenacious, perseverative, ruminative quality of their ideas about body image, weight, and food. We found no obvious differences in attitude, mentation, or cognition between patients with normal or increased CSF AVP levels; but, as noted above, these single unstimulated CSF values may not accurately reflect either the incidence or the severity of the defects. It is also possible that the AVP in CSF significantly influences brain hydration via effects on choroid plexus (Rodriguez and Heller 1970) or capillary permeability (Raichle and Grubb 1978). CSF pressure was normal in our patients; but, to our knowledge, other indices of brain hydration have not been investigated in anorexia nervosa.

Conclusion

These studies have demonstrated that a number of changes occur in neurotransmitter and neuromodulator systems during the syndrome of anorexia nervosa. Whether any of these changes is a primary etiological event that occurs early in the illness or triggers the illness remains an open question. Some of the changes appear to occur predictably in each individual during the stage of illness when the patient is at low weight, such as the decrease in brain dopamine and serotonin metabolism and the increase in CSF total opioids. These are more likely to be compensatory physiological responses to the state of the low weight and low caloric intake. Other changes, such as those in vasopressin function and brain norepinephrine metabolism, are prolonged, last months after nutritional rehabilitation, and are also more heterogeneous between individuals. These neurobiological changes may indicate subtypes of the syndrome with different outcomes or the presence of changes in brain function that may drive a chronic eating disorder. Hopefully these questions may be further explored by following patients longitudinally or by studying acute starvation in normal subjects.

References

Baile CA, Della-Fera MA, McLaughlin CL (1980) Opiate antagonist and agonist and feeding in sheep. Fed Proc 39: 782

Blessing WW, Sved AF, Reis DJ (1982) Destruction of noradrenergic neurons in rabbit brainstem elevates plasma vasopressin, causing hypertension. Science 217: 661–663

Booth DA (1967) Localization of the adrenergic feeding system in the rat diencephalon. Science 158: 515–517

Buijs RM, Swaab J, Dogterom J, van Leeuwen FW (1978) Intra- and extra-hypothalamic vasopressin and oxytocin pathways in the rat. Cell Tissue Res 186: 423–433

Cantwell DP, Sturzenberger KS, Burroughs J, Salkin B, Green JK (1977) Anorexia nervosa, an affective disorder. Arch Gen Psychiatry 34: 1087–1093

Cooper JR, Bloom FE, Roth RH (1982) The biochemical basis of neuropharmacology, 4th edn. Oxford New York

Coppen AJ, Gupta K, Eccleston EG, Wood KM, Arkeling A, de Sousa VFA (1976) Plasma tryptophan in anorexia nervosa. Lancet 1: 961

Curzon G, Joseph MH, Knott PH (1972) Effects of immobilization and food deprivation on the rat brain tryptophan metabolism. J Neurochem 19: 1967–1974

de Wied D (1976) Behavioral effects of intraventricularly administered vasopressin and vasopressin fragments. Life Sci 19: 685–690

Ebert MH, Kartzinel R, Cowdry RW, Goodwin FK (1980) Cerebrospinal fluid amine metabolites and the probenecid test. In: Wood JH (ed) Neurobiology of cerebrospinal fluid. Plenum, New York, pp 97–110

Friedman E, Starr N, Gershon S (1973) Catecholamine synthesis and regulation of food intake in the rat. Life Sci 12: 317–326

Fuxe K, Ungerstedt U (1966) Localization of catecholamine uptake in rat brain after intraventricular injection. Life Sci 5: 1817–1824

Fuxe K, Ungerstedt U (1968) Histochemical studies on the effects of (+)-amphetamine, drugs of imipramine groups, and tryptamine on central catecholamine and 5-hydroxtryptamine neurons

after intraventricular injection of catecholamines and 5-hydroxytryptamine. Eur J Pharmacol 4: 135–144
Gerner RH, Gwirtsman HE (1981) Abnormalities of dexamethasone suppression test and urinary MHPG in anorexia nervosa. Am J Psychiatry 138: 650–653
Gershon ES, Hamovit JR, Schrieber JL, Dibble ED, Kaye W, Nurnberger JI, Andersen A, Ebert M (to be published) Anorexia nervosa and major affective disorders associated in families: a preliminary report. In: Guze SB, Ecols FJ, Barrett JE (eds) Childhood psychopathology and development. Raven, New York
Gold MS, Redmond DE, Kleber HD (1979) Noradrenergic hyperactivity in opiate withdrawal supported by clonidine reversal of opiate withdrawal. Am J Psychiatry 136: 100–102
Gold PW, Weingartner HL, Ballenger JC, Goodwin FK, Post RM (1978) The effects of des-amino-8-d-arginine vasopressin (DDAVP) on behavior and cognition in patients with primary affective disorder. Lancet 2: 992–995
Gold PW, Goodwin FK, Ballenger JC, Robertson GL, Post RM (1981) Central vasopressin function in affective illness. In: de Wied D, Van Keep PA (eds) Hormones and the brain. MTP Press, Brussells, pp 241–253
Gold PW, Kaye W, Robertson G, Ebert M (1983) Abnormalities in plasma and cerebrospinal-fluid arginine vasopressin in patients with anorexia nervosa. N Engl J Med 308: 1117–1123
Goldman HW, Lehr D, Friedman E (1971) Antagonistic effects of alpha- and beta-adrenergically coded hypothalamic neurons on consummatory behavior in the rat. Nature 231: 453–455
Grandison L, Guidotti L (1977) Stimulation of food intake by muscinol and beta endorphin. Neuropharmacology 16: 533–536
Gross HA, Lake CR, Ebert MH, Hiegler MG, Kopin IJ (1979) Catecholamine metabolism in primary anorexia nervosa. J Clin Endocrinol Metab 49: 805–809
Holtzman SG (1974) Behavioral effects of separate and combined administration of naloxone and d-amphetamine. J. Pharmacol Exp Ther 189: 51–60
Jenkins JS, Mather HM, Ang V (1980) Vasopressin in human cerebrospinal fluid. J Clin Endocrinol Metab 50: 364–367
Kantak KM, Wayner MJ, Stein JM (1978) Effects of various periods of food deprivation on serotonin turnover in the lateral hypothalamus. Pharmacol Biochem Behav 9: 529–534
Kaye WH, Pickar D, Naber D, Ebert MH (1982) Cerebrospinal fluid opioid activity in anorexia nervosa. Am J Psychiatry 139: 643–645
Kopin IJ, Gordon EK, Jimerson DC, Polinsky RJ (1982) Relation between plasma and cerebrospinal fluid levels of 3-methoxy-4-hydroxyphenylglycol. Science 219: 73–75
Korf J, Bunney BS, Aghajanian GH (1974) Noradrenergic neurons: morphine inhibition of spontaneous activity. Eur J Pharmacol 25: 165–169
Kron L, Katz JL, Gorzynski G, Weiner H (1978) Hyperactivity, in anorexia nervosa: a fundamental clinical feature. Compr Psychiatry 19: 433–440
Krulich LA, Giachetti A, Marchlewdkako JA, Hefco E, Jameson HE (1977) On the role of the central noradrenergic and dopaminergic systems in the regulation of TSH secretion in the rat. Endocrinology 100: 496–505
Lee PLY (1974) Single-column system for accelerated amino acid analysis of physiological fluids using five lithium buffers. J Biochem Med 10: 107–121
Leibowitz SF (1980) Neurochemical systems of the hypothalamus. In: Morgane PJ, Panksepp J (eds) Handbook of the hypothalamus, vol 3, part a. Dekker, New York, pp 299–437
Leibowitz SF (to be published) Noradrenergic function in the medical hypothalamus: potential relation to anorexia nervosa and bulimia
Loffstrom A (1977) Catecholamine turnover alterations in discrete areas of the median eminence of the 4- and 5-day cyclic rat. Brain Res 120: 113–131
Loullis CC, Felten DL, Shea PA (1979) HPLC determination of biogenic amines in discrete brain areas in food deprived rats. Pharmacol Biochem Behav 1: 89–93
Luerrson TB, Robertson GL (1980) Cerebrospinal fluid vasopressin and vasotocin in health and disease. In: Wood JH (ed) The neurobiology of cerebrospinal fluid. Plenum, New York, pp 613–623
Margules DL (1979) Beta-endorphin and endoloxone: hormones of the autonomic nervous system for the conservation or expenditure of bodily resources and energy in anticipation of famine or feast. Neurosci Biobehav Rev 3: 155–162

Margules DL, Moisset B, Lewrs MJ, Shibuya H, Pert CB (1978) β-Endorphin is associated with overeating in genetically obese mice (ob/ob) and rats (fa/fa). Science 202: 988–991

McCann SM (1970) Neurohormonal correlates of ovulation. Fed Proc 29: 1888–1970

Mecklenberg RS, Loriaux DL, Thompson RH, Anderson AE, Lipsett MB (1974) Hypothalamic dysfunction in patients with anorexia nervosa. Medicine 53: 147–159

Naber D, Pickar D, Dionne RW et al. (1980) Assay of endogenous opiate receptor in human CSF and plasma. Subst Alcohol Actions Misuse 1: 83–91

Naber D, Cohen RM, Pickar D et al. (1981) Episodic secretion of opioid in human plasma and monkey CSF: evidence for a diurnal rhythm. Life Sci 28: 931–935

Perlow M, Ebert M, Gordon E, Ziegler MG, Lake CR, Chase TN (1978) The circadian variation of catecholamine metabolism in the subhuman primate. Brain Res 139: 101–113

Pirke KM, Spyra B (1982) Catecholamine turnover in the brain and the regulation of luteinizing hormone and corticosterone in starved male rats. Acta Endocrinol (Kbh) 100: 168–176

Raichle ME, Grubb RL (1978) Regulation of brain water permeability by centrally released vasopressin. Brain Res 143: 191–194

Reppert SM, Artman HG, Swaminathan S, Fisher DA (1981) Vasopressin exhibits a rhythmic daily pattern in cerebrospinal fluid but not in blood. Science 213: 1256–1259

Ritter RC, Epstein AN (1975) Control of meal size by central noradrenergic action. Proc Natl Acad Sci USA 72: 3740–3743

Ritter S, Wise CD, Steil L (1975) Neurochemical regulation of feeding in the rat. J Comp Physiol Psychol 88: 515–517

Robertson GL (1977) The regulation of vasopressin function in health and disease. Recent Prog Horm Res 33: 333–385

Robertson GL, Shelton RL, Athar S (1976) The osmoregulation of vasopressin. Kidney Int 10: 25–37

Rodriguez EM, Heller H (1970) Antidiuretic activity and ultrastructure of the toad choroid plexus. J Endocrinol 46: 83–91

Rubinow DR, Post RM, Pickar D (to be published) Relationship between urinary free cortisol and CSF opioid binding activity in depressed patients and normal volunteers. Psychol Res

Russell GFM (1967) The nutritional disorder in anorexia nervosa. J Psychosom Res 11: 141–149

Sawyer CH, Hilliard J, Kanematsu S, Scaramuzzi R, Blake CA (1974) Effects of intra-ventricular infusions of norepinephrine and dopamine on LHH release and ovulation in the rabbit. Neuroendocrinology 101: 1064–1070

Scapagnini U, Annunziato L, Clementi G, Di Renzo GF, Schetini G, Fiore Preziosi P (1977) Chronic depletion of brain catecholamines and thyrotropin secretion in the rat. Endocrinology 101: 1064–1070

Slangen JL, Miller NE (1969) Pharmacological tests for the function of hypothalamic norepinephrine in eating behavior. Physiol Behav 8: 885–890

Smythe GA (1977) The role of serotonin and dopamine in hypothalamic-pituitary function. Clin Endocrinol (Oxf) 7: 325–341

Sofroniew MV (1980) Projections from vasopressin, oxytocin, and neurophysin neurons to neural targets in the rat and human. J Histochem Cytochem 28: 475–478

van Wimersma-Greidanus TH, de Wied D (1976) Modulation of passive avoidance behavior of rats by intracerebroventricular administration of antivasopressin serum. Behav Biol 18: 325–333

Vigersky RA, Loriaux DL (1977) Anorexia nervosa as a model of hypothalamic dysfunction. In: Vigersky RA (ed) Anorexia nervosa. Raven, New York, pp 109–122

Vigersky RA, Loriaux DL, Anderson AE, Lipsett ME (1976) Anorexia nervosa: behavioral and hypothalamic aspects. J Clin Endocrinol Metab 42: 517–538

Weil-Malherbe H, Axelrod J, Tomchick R (1959) Blood-brain barrier for adrenaline. Science 129: 1226–1227

Weiner RI, Ganong WH (1978) Role of brain monoamines and histamine in regulation of anterior pituitary secretion. Physiol Rev 58: 905–976

Weingartner H, Gold P, Ballenger JG, Smallberg S, Summers R, Post R, Goodwin FK (1981) Effects of vasopressin on human memory functions. Science 211: 601–603

Wurtman RJ (1967) Effects of nutrients and circulating precursors on the synthesis of brain neurotransmitters. In: Garattini S, Samanin R (eds) Central mechanisms of anorectic drugs. Raven, New York, pp 267–294

Zimmerman EA, Robinson AG (1976) Hypothalamic neurons secreting vasopressin and neurophysin. Kidney Int 10: 12–24

Sleeping and Waking EEG in Anorexia Nervosa

D. J. Kupfer and C. M. Bulik[1]

Introduction

Anorexia nervosa has emerged as a major disease of our current generation. There is evidence that the incidence of the disorder is on the increase (Duddle 1973; Halmi 1974; Jones et al. 1980 Kendell et al. 1973; Theander 1970). Considerable psychological research has extensively mapped and documented familial, sociological, and family correlates of the disorder; several of the more popular treatment modalities have focused on precisely these aspects of anorexia. The search for objective and replicable correlates of the disease has lagged behind the psychologically and sociologically oriented research. However, further elucidation of possible physiological causes and correlates is crucial to a more comprehensive understanding of the disease.

In the search for objective correlates, several scientific approaches have been adopted to examine the etiology and pathophysiology of anorexia nervosa. It has long been suggested that neurophysiological and electrophysiological alterations may be present in anorexia and/or bulimia. To review these issues, we will examine the results and the problems associated with three neurophysiological approaches: the daytime clinical EEG; the all-night sleep EEG; and a combined sleep EEG-neuroendocrine approach. The body of literature concerning daytime EEGs in particular and, to a lesser extent, all-night sleep EEGs, has been hampered by the lack of sound methodological procedures. Problems associated with each of the procedures will be examined separately; however, it is necessary to first delineate some of the difficulties inherent in studying patients suffering from anorexia nervosa.

Different diagnostic criteria have been used (Lundberg and Walinder 1967; Feighner et al. 1972; DSM III). This not only leads to differences in sample characteristics of patients across studies, but also results in the failure to retain homogeneity of the patient sample within individual investigations. This is especially problematic with studies concerning daytime clinical EEGs of anorectic patients. Difficulties arise when investigators attempt to examine daytime EEG abnormalities in patients exhibiting what they generically term "eating disorders". Often subjects exhibiting classic abstaining-type anorexia and patients who exhibit binge- and purge-type tendencies are grouped together under one diagnostic category. Other studies collectively evaluate emaciated, normal weight, and obese binge eaters without recognizing possible differences

---

[1] University of Pittsburgh School of Medicine, Department of Psychiatry, 3811 O'Hara Street, Pittsburgh, PA 15213, USA

in psychological variables, such as distorted body image or intense fear of being overweight in the emaciated group, which would set them aside as a totally separate diagnostic entity. These classification problems represent the most crucial deficiency associated with this body of literature and should be kept in mind when attempting to generalize from or interpret results of investigations in this area.

Daytime Clinical EEGs

A significant amount of attention has been directed towards the examination of daytime clinical EEGs of patients with eating disorders. Studies in this area have been conducted in a wide variety of patients suffering from eating disorders ranging from abstaining anorectics to obese compulsive or binge eaters. The major problems associated with this body of literature arise when subject groups are not separated according to clinical symptoms and with "overzealous" interpretation of the resultant clinical EEGs.

Three separate categories of allegedly "abnormal" clinical EEG patterns appear to be associated with eating disorders. The first category (Table 1) encompasses abnormalities in background activity. A generalized slowing in background activity is the most commonly reported finding under this heading. Neil et al. (1980) estimate that 34.5% of anorectics display some sort of abnormally slow background activity, and Crisp et al. increase this estimate to 59% of anorectics. Although this slowing of background activity is seen in a large percentage of anorectic patients, it would be incorrect to define the phenomenon as a marker of anorexia per se. The slowing of background activity could well be reflective of the hormonal or electrolyte changes associated with either rapid weight loss or chronic low body weight. The lack of reliable data on the effects of rapid weight loss on the clinical EEGs of normals contributes to the difficulty of properly interpreting the EEG results of anorectic patients. Slow background

Table 1. Clinical EEG findings in anorexia nervosa. (Adapted from Neil et al. 1980)

EEG findings	Anorexia nervosa subgroups		
	Primary ($n = 36$)	Secondary ($n = 19$)	Total ($n = 55$)
Normal	20 (55.6%)	12 (63.2%)	32 (58.2%)
Abnormal (total)[a]	16 (44.4%)	7 (36.8%)	23 (41.8%)
EEG slowing	9	10	19
Sharp wave/spike transients	5	3	7
Generalized spike/wave	1	1	2
Focal spikes/spike-wave	3	1	4
Mitten patterns	1	1	2
Extreme spindles	2	1	3
Unstable hyperventilation response	5	3	8

[a] $\chi^2 = 0.206$, 1 d.f., difference not statistically significant

activity is rare in the waking EEGs; however, it is unknown whether this figure rises when normals undergo rapid weight loss.

The second category of abnormalities concerns more specific EEG patterns, namely 14- and 6-per-second positive spikes, 6-per-second spike and wave activity (or phantoms), and B-mitten patterns. Evidence of 14- and 6-per-second positive spikes was found in abstaining anorectics (Gibbs and Gibbs 1964; Shimoda and Kitagawa 1973) and compulsive or binge eaters (Green and Rau 1974; Rau and Green 1975; Rau et al. 1979). Six-per-second spike and wave activity or phantoms were evidenced in abstaining anorectics, bulimics, and binge or compulsive eaters of various body weights ranging from emaciated to obese (Crisp et al. 1968; Lundberg and Walinder 1967; Shimoda and Kitagawa 1973; Davis et al. 1974). B-mitten patterns, although less common, were viewed in anorectic patients (Neil et al. 1980; Gibbs and Gibbs 1964) and binge eaters (Rau et al. 1979).

The third category of clinical EEG abnormalities concerns unstable or prolonged hyperventilation responses. Crisp et al. (1968) defined abnormal hyperventilation response as runs of high-voltage slow waves consistently evident in patients over 25 or for greater than 60 s after the cessation of hyperventilation in younger patients. They found that 31% of anorectic patients and 9% of controls showed unstable hyperventilation responses. Neil et al. (1980) claim that 14.6% of anorectic patients demonstrated exaggerated or prolonged responses to hyperventilation; unfortunately, these authors provided neither the criterion used to define abnormal response nor comparative data for normal.

Investigators have taken various degrees of liberty in interpreting the occurrences of these three categories of clinical EEG abnormalities. They have been collectively interpreted as being indicative of organic cerebral lesions (Lundberg and Walinder 1967), thalamic or hypothalamic disturbances (Gibbs and Gibbs 1964), brain stem dysfunction or correlates of seizure activity (Hughes et al. 1965), or a more general disordered neurophysiological substrate that may contribute to eating disorders (Rau et al. 1979).

It is highly probable that investigators have greatly overinterpreted these clinical EEG findings and that the abnormalities observed are more accurately described as by-products of the metabolic or endocrine disturbances that accompany either drastic weight loss, chronic low body weight, or binge and purge activities, than as accurate indicators of underlying brain pathologies that are correlates of eating disturbances. Evidence for this claim is provided by Crisp et al. (1968), who correlated incidences of EEG abnormalities (Table 2) with various clinical measurements associated with anorexia and found that the slowing of background activity was correlated with duration of illness, extremely high or extremely low resting pulse rates, and low levels of serum sodium potassium and chloride. Unstable hyperventilation responses correlated with duration of illness, low blood sugar, high plasma insulin, and low serum sodium and chloride levels. The investigators claim that patients with a duration of illness greater than 5 years who tend to binge and purge often have more severely disturbed serum electrolyte values and consequently display a greater amount of EEG abnormalities.

Table 2. Relationships between daytime EEG findings and various physical and biochemical measurements. (Adapted from Grisp et al. 1968)

		Daytime EEG Normal	Daytime EEG Abnormal	P
EEG background activity	Duration of illness < 5 years	12	10	0.002
	Duration of illness ≥ 5 years	1	9	
Resting pulse	55–80/min (inclusive)	8	6	0.033
	< 55 or > 80/min	3	13	
Blood sugar levels	> 70 mg/100 ml	4	4	0.334
	70 mg/100 ml or less	6	13	
Plasma insulin levels	≥ 20 microunits	2	3	0.334
	< 20 microunits	5	3	
Serum sodium levels	≥ 135 mEq/l	11	12	0.019
	< 135 mEq/l	0	6	
Serum potassium levels	≥ 3.9 mEq/l	9	6	0.014
	< 3.9 mEq/l	2	12	
Serum chloride levels	≥ 95 mEq/l	10	9	0.007
	< 95 mEq/l	0	9	

With specific EEG abnormalities (14- and 6-per-second positive spikes, 6-per-second spike and wave activity and B-mitten patterns), it should be realized that the daytime EEG is a sensitive but nonspecific indicator and that EEG abnormalities similar to those observed in some anorectics would also be found in patients with other disorders and in a certain percentage of normals. Maulsby (1979) classifies the 14- and 6-per-second positive spike and the 6-per-second spike and wave activity as EEG patterns of uncertain diagnostic significance. The 14- and 6-per-second positive spike, he claims, has been correlated with a heterogeneous collection of complaints, primarily in teenagers and children. Indeed, Maulsby suggested that within the next 5 years the 14- and 6-per-second positive spike would be classified under normal EEG patterns. Given that the anorectic patients studied were primarily in their teens, it is understandable why this pattern, frequently seen in teenagers, emerged in the anorectic sample.

Maulsby also places the 6-per-second spike and wave pattern or phantom under the category of patterns of uncertain diagnostic significance. Thomas and Klass (1968) claim that the 6-per-second pattern is typical for the 15- to 19-year-old age group and that it provides no proof for the presence of epilepsy. It is evident that the attempt to include compulsive eating under the heading of seizure disorders is largely unjustified on the basis of the clinical EEG data above.

Thus, the direction further research in this area should follow is clear. The pattern to date has been to perform the clinical EEGs and then to speculate as to the meaning of the abnormalities found. First, definitive studies should be conducted determining the effects of weight loss or electrolyte disturbance on the EEGs of normals to determine whether the abnormalities seen in anorectics are indeed correlates of nutritional or body weight state as opposed to trait characteristics of patients predisposed to eating disorders.

This could be further examined via a high-risk, long-term study. The social, familial, and personality characteristics of patients with anorexia have been thoroughly researched by several investigators. An examination of high-risk subjects would provide more information concerning any premorbid EEG abnormalities that may be indicative of a predisposition to anorexia nervosa. Continuing the study through emergent anorectic periods would then provide information on state characteristics of EEG abnormalities associated with both acute and chronic stages of anorexia.

The relative lack of specificity in the routine clinical EEG must be evaluated in determining the worth of the information accrued using this tool. Green and Rau (1974) caution that an initial or single normal EEG does not preclude a possible and significant periodic EEG abnormality. Duplicate or repeat EEGs are often lacking in the majority of studies in this area. Perhaps the clinical EEG will eventually be abandoned for more specific and accurate measures of brain activity (e.g., evoked potentials). For example, the research of Leibowitz's group (Leibowitz and Brown 1980a) has pinpointed the perifornical region of the hypothalamus as the feeding inhibition center of the rat brain and they have mapped pathways contributing to both feeding stimulation and feeding inhibition (Leibowitz and Roussakis 1979a, b; Leibowitz and Brown 1980a, b). Analogous information for the human brain and more sensitive research techniques would minimize the speculative nature of the hypotheses introduced on the basis of EEG data and would provide more specific information with regard to the possible neurophysiological targets that may be involved in eating disorders.

All-Night Sleep EEGs

When examining the all-night sleep EEG studies of patients with anorexia, one is immediately struck by the paucity of research. Other than studies conducted by Crisp et al. and in this country by our research group, little has been undertaken to examine possible sleep disturbances in patients suffering from anorexia nervosa.

Sleep measures can be roughly categorized under the headings of sleep continuity variables (sleep onset difficulty, awakenings, and sleep maintenance), sleep architecture measures (Table 3) (specific percentages of sleep stages), and REM sleep measures. Disturbances in all three of these categories have been reported in the sleep of anorectic patients; however, the paucity of research and methodological shortcomings limit the number of conclusions one can draw from the extant research. Once again, the problems with consistency of diagnostic and inclusion criteria arise. Lacey et al. (1975) state only that their patients exhibited "unequivocal evidence of primary anorexia nervosa", which hardly provides a clear definition of the subject population. The 1975 paper reporting this study

Table 3. Glossary of EEG sleep variables

Term	Explanation
Sleep latency	Time from lights out until the appearance of stage 2 sleep
Early morning awakening	Time spent awake from the final awakening until the subject gets out of bed
Awake	Time spent awake after sleep onset and before the final awakening in the morning
Time spent asleep (TSA)	Time spent asleep less any awake time during the night after sleep onset
Awake/TSA	Percentage of time awake over time spent asleep
TSA/total recording period (sleep efficiency)	Ratio of time spent asleep to total recording period
Sleep architecture	Various percentages of each stage of sleep: stage 1, stage 2, stage 3, and stage 4 (stages 3 and 4 are combined to yield delta sleep), stage 1 REM percentage and stage 2 REM percentage (the portion of the night when REMs are present during stage 1 and stage 2 sleep, respectively)
REM latency	Number of minutes asleep until the onset of the first REM period
REM activity (RA)	Each minute of REM sleep is rated on a 9-point scale (0–8) for the relative amount of REM occurring; the sum for the whole night provides the REM activity
RA/TSA (average REM activity)	Ratio of REM activity to time spent asleep
RA/REM time (REM density)	Ratio of REM activity to REM time

contains the following criteria: 1) history of weight loss to at least 20% less than the matched population mean; 2) all females displaying amenorrhea for at least 6 months or never having menstruated; and 3) all characteristic psychopathology of anorexia nervosa including fear of normal adolescent weight and excessive concern with being fat.

Our reported studies have used RDC or Feighner criteria for the diagnosis of anorexia nervosa. The problem with comparing results across these studies, given the disparity in diagnostic criteria, is apparent. The criteria used by Lacey et al. (1975) tend to be overinclusive. They fail to examine weight loss relative to premorbid body weight, the psychological and attitudinal manifestations of the fear of gaining weight or being fat, and the additional clinical symptoms referenced in category F of Feighner's criteria.

Keeping these differences in mind, the first category of sleep to be examined contains the sleep continuity variables. Crisp et al. (1967, 1970) claimed that anorectics suffered from insomnia or poor sleep which manifested itself primarily in early morning awakening or frequent awakening during the middle of the night. Crisp (1967) demonstrated that the early morning awakening in anorectics was not related to mood or any of the clinical features of the disorder. The 1970 study by Crisp et al. claimed that early morning awakening is more closely associated with disturbances and changes in nutritional status than with changes and disturbances in emotional state. In addition, Lacey et al. (1975,

1976) claim that these sleep disturbances are reversible upon restoration of normal body weight.

The 1975 study conducted by Lacey et al. examined the EEG sleep of ten anorectics upon admission and after restoration of target weights as determined by actuarial charts from the general population. In terms of sleep continuity variables, they found that anorectics spent less time asleep at low body weights, that after resumption of target body weight there was a significant increase in total sleep time (TSA + awake time), and that there was decreased wakefulness during the refeeding sleep study. In addition, they found that wakefulness at low body weights was concentrated toward the end of the sleep period. Unfortunately, they provide no comparative data from normals to determine whether the findings at admission were significantly different from normal controls.

Foster et al. (1976) provide some evidence for differences between normal and anorectic sleep in a study of five drug-free anorectics and five normal controls who were sleep-studied at admission and upon reattainment of 75% of predicted ideal body weight. They found no significant differences between normals and anorectics at admission. The only differences evidenced were in REM variables, which will be discussed later. At minimal refeeding weight, anorectics displayed a greater amount of nocturnal awakening than normal controls. However, the distribution of this nocturnal awakening throughout the night was not provided.

The second area of sleep variables concerns the basic architecture of sleep. There appear to be definite changes in sleep architecture when normal body weight is regained after an anorectic episode. In terms of sleep architecture variables, Crisp et al. (1970) found significant increases in stages 3 and 4 and REM sleep and significant decreases in wakefulness and stage 1 once patients had regained their target weights.

The 1975 study conducted by Lacey et al. confirmed these findings and presented additional information regarding the progression of changes in sleep architecture as target weights were gradually regained. They examined this by sleep-studying six of their ten subjects on three occasions: at admission, when they were within 15% of their target weights, and once target weights were regained. Along with the aforementioned increases in total sleep time, they found increases in slow-wave sleep (SWS) and REM sleep and decreases in wakefulness once target weights were attained.

A comparison of the pretreatment sleep studies and those performed when the patients were 15% below target weight and at target weight revealed that EEG sleep architecture changes may occur in two distinct phases. After the initial weight gain, TSA, WSW, and REM all increased. Upon final weight gain to target weights, a hypothesized shift in metabolism was reflected in further increases in REM and a decrease and eventual stabilization of SWS. Confirmation of this theory warrants closer attention and, if validated, would shed further light on the ongoing controversy concerning the role of SWS in the process of body restoration. If SWS does indeed play a role in the restoration or recovery of the body, then the temporary increase in SWS directly after the acute anorectic episode could be a reaction to demands of a body depleted by drastic weight loss or a chronic underweight condition. Then as body weight increases

Table 4. EEG sleep parameters and waking EEG findings (Adapted from Neil et al. 1980)

EEG sleep parameters	Anorexia nervosa patients		Normal controls ($n = 10$)
	Normal EEG ($n = 10$)	Abnormal EEG ($n = 7$)	
Sleep continuity			
Sleep latency (min)	17.8 ± 2.3	32.2 ± 4.3	27.2 ± 3.0
Time spent asleep (min)	356.5 ± 11.7	319.6 ± 12.6	371.1 ± 7.8
Intermittent awakening (%)	7.4 ± 5.8	6.6 ± 4.2	0.4 ± 0.4
Sleep efficiency (%)	90.1 ± 2.9	86.1 ± 3.8	92.0 ± 3.5
Sleep architecture			
Stage 1 (%)	7.4 ± 0.7	6.6 ± 0.5	6.2 ± 1.4
Stage 2 (%)	66.1 ± 2.8	67.5 ± 5.6	58.9 ± 3.2
Delta (%)	1.8 ± 1.3[a,f]	12.6 ± 4.9	13.5 ± 3.8
Stage 1 REM (%)	22.4 ± 2.2	13.2 ± 2.4[b,c]	20.3 ± 2.4
Stage 2 REM (%)	3.0 ± 1.4	2.1 ± 1.3	0.8 ± 0.2
Selected REM measurements			
REM latency (min)	62.1 ± 7.1[e]	67.8 ± 14.9	90.8 ± 7.6
REM activity (U)	94.8 ± 19.9[e]	41.1 ± 9.4[a,d]	123.9 ± 14.4
REM activity/total sleep	0.26 ± 0.05[e]	0.12 ± 0.03[d]	0.33 ± 0.03
REM density (RA/RT)	1.11 ± 0.18[e]	0.93 ± 0.12[c]	1.65 ± 0.20

[a] $P < 0.05$
[b] $P < 0.01$, normal vs abnormal EEG
[c] $P < 0.05$
[d] $P < 0.01$, abnormal EEG vs control
[e] $P < 0.05$
[f] $P < 0.01$, normal EEG vs control

and approaches normal and damage done to the body through starvation is gradually repaired, the demand for SWS diminishes and the percentage of SWS will gradually decline and stabilize. This interpretation is, however, purely speculative; it must first be determined whether this "two-phase" recovery phenomenon does indeed occur and whether SWS is in any way implicated in the body's restorative process.

Studies performed by Neil et al. (1980) and Foster et al. (1976) provide less convincing evidence for drastic changes in sleep architecture associated with anorexia nervosa. Neil et al. (1980), who divided patients according to whether they exhibited normal or abnormal daytime EEGs, found a greater reduction (Table 4) in the percentage of REM sleep in the group of anorectics displaying "abnormal" EEGs and a significantly lower amount of delta sleep in patients exhibiting normal daytime EEGs. First, on the basis of the previous discussion, it is unclear whether the division on the basis of daytime EEG "abnormalities" is justified. Second, the investigators fail to provide data on changes that occur upon reattainment of normal target weights.

The other study from our research group (Foster et al. 1976) revealed no significant differences in sleep architecture variables between admission sleep studies and studies performed at 75% of predicted ideal body weight. They did, however, find a greater amount of stage 1 sleep in anorectics at minimal refeeding weight than in normal controls. One possible reason for the lack of significant findings in sleep architecture variables between admission and refeeding sleep studies could be that at the time of the final study patients had only attained 75% of their predicted ideal body weight. The theory of Lacey et al. (1975) in this situation implies that the final changes in sleep had not yet been attained because the patients had not yet attained their ideal body weight. However, according to this theory, an increased amount of SWS would be expected at the time of the final sleep study.

A true examination of changes in delta sleep, however, would not only involve the recording of amount and percentage of delta sleep, but would examine possible changes in SWS density throughout the course of the disorder and upon recovery. Even in the absence of an increase in amount or percentage of SWS in relation to overall sleep architecture, Lacey's hypothesis may be reflected in an increase in SWS density, meaning an increase in the actual delta-wave count during periods of SWS. This possibility should be investigated with computer applications of either a delta-wave analyzer or spectral analysis before any theories about SWS changes associated with the acute phase of anorexia nervosa or recovery from the disorder are dismissed.

The final aspect of sleep to be examined concerns more detailed analyses of phasic REM activities. The studies conducted by Lacey et al. fail to address these questions. Although they do find increases in REM sleep time upon refeeding, even when corrected for increased sleep time during the later portion of the night, they fail to examine measures such as total REM activity, average REM activity, or REM density. The information available on these parameters comes exclusively from the studies by Neil et al. (1980) and Foster et al. (1976). The study by Foster et al. confirmed these results and found, along with a lower percentage of REM sleep, lower total REM activity and lower average REM activity in anorectics than in normals at the time of admission. At minimal refeeding weight, anorectics continued to show lower total REM activity and lower average REM activity than normals. Similarly, Neil et al. found reduced REM activity and reduced REM density (ratio of total REM activity to total REM time) in anorectics and found the reduction in REM activity and REM percentage to be greatest in the group of anorectics displaying abnormal clinical daytime EEGs.

Further investigation into the changes in phasic REM parameters may be the key to determining whether an underlying medical/neurological disturbance is present in the disorder. Reductions in REM density as a biological marker have been investigated by King et al. (1981) in a study examining patients with 1) CNS disorders; 2) medical disorders; 3) primary affective disorders only; and 4) other psychiatric disorders, compared with normal controls. Granted, the group divisions retained a large degree of heterogeneity within each category, and, in addition, no anorectics were included in the sample; however, low REM density was most consistently associated with patients exhibiting medical/neurological disorders. Although replication is necessary, the further examination of phasic REM disturbances may be crucial to understanding the etiology of anorexia nervosa. In addition, these measures may prove to be predispositional or trait measures of the disorder, whereas changes in sleep continuity and sleep architecture variables may simply be reflective of the state of maintaining a chronically underweight level.

The reversibility of all the aforementioned changes remains unclear when the results are reviewed across studies. Crisp et al. (1970), who found a reduction in REM time during acute stages of anorexia, claim that this reduction is reversed upon reattainment of normal weight; however, as previously mentioned, no data were presented to examine whether changes in phasic REM parameters undergo a parallel reversal. Foster and Kupfer (1976) have provided very preliminary evidence which seems to indicate that phasic REM parameters, which were found to be reduced in acute phases of anorexia, may be more inclined to remain low even after weight gain. Three sisters, monozygotic twins concordant for anorexia and a sister exhibiting hyperbulimia with fluctuating weight, displayed consistent reductions in REM time and phasic REM activity whether they were studied during active periods of anorexia, anorexia in clinical remission, or active bulimia. Although more extensive and comprehensively designed studies with a greater number of subjects are required, this preliminary evidence offers a guideline for future research to determine whether phasic and possibly even tonic REM disturbances may be genotypic markers for an underlying CNS disturbance resulting in a predisposition to eating disturbances.

Although all-night EEG studies in patients with anorexia nervosa are far from exhaustive, it is apparent that documentable alterations in REM parameters do accompany at least the acute phase of anorexia nervosa. Several investigators have attempted to relate anorexia nervosa to affective disorders on the basis of similarities in sleep disturbances and neuroendocrine anomalies associated with the two disorders. Currently, in affective disorder research, a combined approach of all-night sleep EEG and neuroendocrine sampling is being used to determine the relationship between endocrinological and sleep patterns. Given that both sleep and abnormalities in neuroendocrine parameters have been implicated in anorexia nervosa, the adaptation of the sleep neuroendocrine profile approach appears to be warranted in the study of anorexia.

A brief comparison of sleep and neuroendocrine profiles of patients with affective disorders and anorexia nervosa reveals a significant amount of overlap on both axes. Characteristic profiles of primary endogenous depressives include elevated plasma cortisol levels, nonsuppression or early escape from the dexamethasone suppression test, decreased REM latency, early morning awakening, increased intermittent awakening, increased REM sleep especially during the second half of the night, and decreased amounts of SWS. Anorectic patients, in comparison, have consistently demonstrated elevated plasma cortisol levels (Gerner and Wilkins 1983; Halmi et al. 1978; Doerr et al. 1980; Walsh et al. 1981; Casper et al. 1979; Boyar et al. 1977), nonsuppression of cortisol in the dexamethasone suppression test with a tendency towards early escape as higher body weights are reached (Gerner and Gwirtsman 1981), and a prolonged cortisol half-life. Boyar et al. (1977) have indicated that the normal circadian variation of cortisol remains intact but at consistently elevated levels. In attempting to relate anorexia and affective disorders, Winokur et al. (1980) and Cantwell et al. (1977) have demonstrated a significantly higher incidence of

Table 5. Distribution of primary affective disorder (PAD) in relatives of anorectic patients and controls. (Adapted from Winokur et al. 1980)

Number of family members with PAD	Anorectic families		Control families	
	n	%	n	%
0	6	24	13	53
1	7	28	7	28
2 or more	12	48[a]	5	20

[a] Significantly different from controls ($\chi^2 = 5.46$, $P < 0.025$)

affective disorders in first-degree relatives of anorectic patients than in those of normal controls (Table 5). Finally, a large percentage of anorectics suffer from depressive episodes during the acute phase of anorexia and, after recovery from anorexia, a significant percentage of anorectics return for treatment of depression.

A recent study performed by Jarrett et al. (1983) demonstrated that the nadir of nocturnal plasma cortisol concentration was significantly higher in depressives and that the characteristic nocturnal rise in plasma cortisol occurred significantly closer to sleep onset in depressives than in normal controls. This line of investigation should also be pursued with anorectic patients from several experimental angles. Sleep and cortisol secretory patterns of anorectics should be compared with those in both normal controls and depressives and, within the anorectic sample, a comparison should be made between anorectics who display depressive symptomatology and those who do not. This approach would provide further information regarding the relationship between anorexia nervosa and affective disorders and would extend the information base on biological correlates of anorexia nervosa.

In addition to the sleep-neuroendocrine approach, long-term follow-up studies are essential to the understanding of the correlates of anorexia. It is as yet unexplained whether the cortisol elevation, DST nonsuppression, and alterations in REM sleep exist premorbidly and remain postmorbidly. Although there is some evidence to indicate that the cortisol and tonic REM variables may be state markers of anorexia, more extensive investigations should be undertaken to examine predispositional variables, recovery indicators, and variables which may be predictive of relapse.

In comparison with state of the art research in endogenous depression where specific sleep and neuroendocrine abnormalities are rapidly being recognized as biological markers of that disorder, the sleep and sleep-neuroendocrine patterns associated with anorexia nervosa fail to demarcate a coherent biological picture. One fundamental assumption underlying current research is that the syndrome of anorexia nervosa represents a homogeneous diagnostic entity. If this assumption is true, one would expect a coherent biological picture among anorectic patients. Without the uniformity of biological findings across patients, one must question the fundamental assumption of anorexia nervosa as a homogeneous entity. The familiar cluster of symptoms associated with anorexia nervosa (emaciation, weight loss, amenorrhea, distorted body image, etc.) may create a false impression of homogeneity of these patients, while that symptom cluster may be the end point of disorders of diverse etiologies which result in the same clinical picture.

Various theories on the etiology of anorexia provide possibilities for hypothesizing what these diverse etiologies or subclassifications may be. The psychobiological theory points to the possibility of a subgroup of anorectics whose emaciation is due primarily to disturbances of eating or satiety centers of the hypothalamus. Studies by Winokur and Cantwell revealing the high incidence of affective disorders in anorectics and studies showing similarities in the response to the dexamethasone suppression test and in cortisol levels (Gerner and Gwirtsman 1981) may indicate that there is a specific subgroup of anorectics in whom the disorder is coupled with depressive symptomatology. A third category, as suggested by Bruch (1962), emphasized the delusional misconceptions in body images and may delimit a specific subgroup of anorectics who suffer primarily from a thought disorder. The goal of future research would then be to first divide patients along the guidelines of these subclassifications and then examine biological and psychological correlates of each of the subsamples. Such an approach would allow a finer discrimination among patients with anorectic symptoms and would enhance the chances of accurate diagnosis and successful treatment for patients suffering from the disorder.

Acknowledgements. This work was supported in part by National Institute of Mental Health grants MH # 30915 and MH # 24652 and by a grant from the John D. and Catherine T. MacArthur Foundation Research Network on the Psychobiology of Depression.

References

Boyar RM, Hellman LD, Roffwarg H, Katz J, Zumoff B, O'Connor J, Bradlow L, Fukushima DK (1977) Cortisol secretion and metabolism in anorexia nervosa. N Engl J Med 269: 190–193

Bruch H (1962) Perceptual and conceptual disturbances in anorexia nervosa. Psychosom Med 24: 187–194

Cantwell DP, Sturzenberger S, Burroughs J, Salkin B, Green JK (1977) Anorexia nervosa: an affective disorder? Arch Gen Psychiatry 34: 1087–1093

Casper RC, Chatterhorn R, Davis JM (1979) Alterations in serum cortisol and its binding characteristics in anorexia nervosa. J Clin Endocrinol Metab 49: 406–411

Crisp AH (1965a) Clinical and therapeutic aspects of anorexia nervosa. A study of 30 cases. J Psychosom Res 9: 67–78

Crisp AH (1965b) Some aspects of the evaluation, presentation and follow-up of anorexia nervosa. Proc R Soc Med 58: 814–820

Crisp AH (1967) The possible significance of some behavioral correlates of weight and carbohydrate intake. J Psychosom Res 11: 117–131

Crisp AH, Fenton GW, Scotton L (1967) The electroencephalogram in anorexia nervosa. EEG Clin Neurophysiol 23: 490

Crisp AH, Fenton GW, Scotton L (1968) A controlled study of the EEG in anorexia nervosa. Br J Psychiatry 114: 1149–1160

Crisp AH, Stonehill E, Fenton GW (1970) An aspect of the biological basis of the mind-body apparatus: the relationship between sleep, nutritional state and mood in disorders of weight. Psychother Psychosom 18: 161–175

Davis KL, Qualls B, Hollister L, Stunkard AJ (1974) EEG's of binge eaters. Am J Psychiatry 131: 1409
Doerr P, Fichter M, Pirke KM, Lund R (1980) Relationship between weight gain and hypothalamic pituitary adrenal function in patients with anorexia nervosa. J Steroid Biochem 13: 529−537
Duddle M (1973) An increase of anorexia nervosa in a university population. Br J Psychiatry 123: 711−712
Foster FG, Kupfer DJ (1976) REM sleep in a family with anorexia nervosa. Sleep Res 5: 141
Foster FG, Kupfer DJ, Spiker DJ, Grau T, Coble PA, McPartland RJ (1976) EEG sleep in anorexia nervosa. Sleep Res 5: 143
Feighner JP, Robins E, Guze S (1972) Diagnostic criteria for use in psychiatric research. Arch Gen Psychiatry 26: 57−63
Gerner RH, Gwirtsman HE (1981) Abnormalities of dexamethasone suppression test and urinary MHPG in anorexia nervosa. Am J Psychiatry 138: 650−653
Gerner RH, Wilkins JN (1983) CSF cortisol in patients with depression mania or anorexia nervosa and in normal subjects. Am J Psychiatry 140: 92−94
Gibbs FA, Gibbs E (1964) Atlas of electroencephalography, vol III. Addison-Wesley, London
Green RS, Rau JH (1974) Treatment of compulsive eating disturbances with anticonvulsant medication. Am J Psychiatry 131: 428−432
Halmi KA (1974) Anorexia nervosa: demographic and clinical features in 94 cases. Psychosom Med 36: 18−26
Halmi KA, Dekirmenjian H, Davis JM, Casper R, Goldberg S (1978) Catecholamine metabolism in anorexia nervosa. Arch Gen Psychiatry 35: 458−460
Hughes JR, Schlagenhauff JE, Mayoss M (1965) Electroclinical correlations in six per second spike and wave complex. EEG Clin Neurophysiol 18: 71−77
Jarrett DB, Coble PA, Kupfer DJ (1983) Reduced cortisol latency in depressive illness. Arch Gen Psychiatry 40: 506−511
Jones DJ, Fox MM, Barbigan HM, Hutton HE (1980) Epidemiology of anorexia nervosa in Monroe County, NY 1960−1976. Psychosom Med 42: 551−558
Kendell RE, Hali DJ, Hailey A, Babigan HM (1973) The epidemiology of anorexia nervosa. Psychol Med 3: 200−203
King D, Akiskal HS, Lemmi H, Wilson W, Belluomini J, Yerevanian RI (1981) REM density in the differential diagnosis of psychiatric from medical-neurological disorders: a replication. Psychiatry Res 5: 267−276
Klass DW, Daly DD (eds) (1979) Current practice of clinical electroencephalography. Raven, New York
Lacey JH, Crisp AH, Kalucy RS, Hartmann MK, Chen CN (1975) Weight gain and the sleep electroencephalogram: a study of ten patients with anorexia nervosa. Br Med J 4: 556−558
Lacey JH, Crisp AH, Kalucy RS, Hartmann MK, Chen CN (1976) Study of EEG sleep characteristics in patients with anorexia nervosa before and after restoration of matched population mean weight consequent of ingestion of a "normal diet". Postgrad Med J 52: 45−49
Leibowitz SF, Brown L (1980a) Histochemical and pharmacological analysis of noradrenergic projects to the paraventricular hypothalamus in relation to feeding stimulation. Brain Res 201: 289−314
Leibowitz SF, Brown L (1980b) Histochemical and pharmacological analysis of catecholaminergic projections to the perifornical hypothalamus in relation to feeding inhibition. Brain Res 201: 315−345
Leibowitz SF, Rossakis C (1979a) Mapping study of brain dopamine and epinephrine sensitive sites which cause feeding suppression in the rat. Brain Res 172: 101−113
Leibowitz SF, Rossakis C (1979b) Pharmacological characterizations of perifornicap hypothalamic dopamine receptors mediating feeding inhibition in the rat. Brain Res 172: 115−130
Lundberg O, Walinder J (1967) Anorexia nervosa and signs of brain damage. Int J Neuropsychiatry 3: 165−173
Maulsby RL (1979) EEG patterns of uncertain diagnostic significance. In: Klass DW, Daly DD (eds) Current practice of clinical electroencephalography. Raven, New York

Neil JF, Merikangas JR, Foster FG, Merikangas KR, Spiker DG, Kupfer DJ (1980) Waking and all-night sleep EEG's in anorexia nervosa. Clin Electroencephalogr 11: 9–15

Rau JH, Green RS (1975) Compulsive eating: a neuropsychologic approach to certain eating disorders. Compr Psychiatry 16: 223–231

Rau JH, Struve FA, Green RS (1979) Electroencephalographic correlates of compulsive eating. Clin Electroencephalogr 10: 180–189

Shimoda Y, Kitagawa T (1973) Clinical and EEG studies on the emaciation (anorexia nervosa) due to disturbed function of the brain stem. J Neural Transm 34: 195–204

Theander S (1970) Anorexia nervosa: a psychiatric investigation of 94 female patients. Acta Psychiatr Scand [Suppl] 214: 24–31

Thomas JE, Klass DW (1968) Six per second spike wave patterns in electroencephalogram. Neurology 18: 587–593

Walsh T, Katz JL, Levin J, Kream J, Fukushima DK, Weiner H, Zumoff B (1981) The production rate of cortisol declines during recovery from anorexia nervosa. J Clin Endocrinol Metab 53: 203–205

Winokur A, March V, Mendels J (1980) Primary affective disorder in relatives of patients with anorexia nervosa. Am J Psychiatry 137: 695–698

Gastric Function in Primary Anorexia Nervosa[1]

A. Dubois[2], H. A. Gross[3], and M. H. Ebert[3]

Introduction

Gastrointestinal function plays an important role in the pathophysiology of starvation as well as in that of primary anorexia nervosa (PAN). Acute gastric dilatation has been reported in PAN patients (Evans 1968; Jennings and Klidjian 1974; Russell 1966; Scobie 1973) as well as in emaciated prisoners of war during refeeding after prolonged starvation (Markowski 1947). In another study, dilatation of the proximal duodenal loop without delay in gastric emptying was demonstrated in 10 of 20 anorexia nervosa patients undergoing barium meal X-ray examination (Scobie 1973). Until recently, the only information available concerning gastric function in relation to nutritional status in man was that gastric emptying was slowed in normal volunteers during starvation (Keys et al. 1950).

In the studies summarized below, we measured gastric emptying and gastric output concurrently in a group of patients with PAN, before and after weight gain, and compared the results with those obtained in 11 healthy controls (Dubois et al. 1979). In addition, we attempted to treat the gastric abnormality with a cholinergic agent (Dubois et al. 1981).

Patients

Fifteen female patients with PAN and 11 healthy controls in the same age range were the subjects of this study. The PAN patients and the healthy controls were admitted to the Clinical Center at the National Institutes of Health. The 11 healthy controls consisted of eight males and three females and ranged in age from 20 to 31, with a mean age of 23 years. Their weight was 68 ± 3 kg (mean ± SEM). They had no medical or psychiatric illness and had no previous history of gastrointestinal disease. The 15 PAN patients ranged in age from 14 to 32, with a mean age of 24 years. Their weight was 34 ± 1 kg. Upon admission, a thorough medical and psychiatric evaluation demonstrated that they met specific criteria for the illness (Feighner et al. 1972). After 10 days of evaluation during which gastric analysis was performed, the patients were treated over the next 4–10 weeks with behavior modification therapy; tube feeding was occasionally required to maintain food intake. During the last week of treatment, six of the 15 patients were again studied after a weight gain of 11 ± 3 kg. At that time, their supine and standing blood pressure and pulse rate were not significantly different from values observed in healthy controls.

---

1 The opinions and assertions contained herein are the private views of the authors and are not to be construed as official or as reflecting the views of the Department of Defense
2 Department of Medicine, Uniformed Services University of the Health Sciences, 4301 Jones Bridge Road, and
3 Laboratory of Clinical Sciences, National Institute of Mental Health, Bethesda, MD 20814, USA

The Psychobiology of Anorexia Nervosa
Edited by K. M. Pirke and D. Ploog
© Springer-Verlag Berlin Heidelberg 1984

Methods

Gastric acid output, fluid output, intragastric volumes, and fractional emptying rates were determined in each subject using a dye dilution technique previously described and validated (Dubois et al. 1977).

Values obtained during the second to fourth 10-min fasting intervals were averaged for each individual, and the mean ($\pm$ SEM) fasting values were calculated for each group. Responses to the distension stimulation were calculated as the mean acid output, fluid output, and fractional emptying rate observed within 60 min following the water load. Maximal responses to the continuous humoral stimulation were calculated as the mean of the two acid output, fluid output, and fractional emptying rates determined 40–60 min following the start of pentagastrin infusion (6 µg · kg^{-1} · h^{-1}; steady-state values of maximal acid output).

Since fractional emptying rates for one of the patients (K. H.) were outside the confidence limits (set as the mean $\pm$ 2 SD) for 15 PAN patients, results were analyzed without K. H.

Five of the female patients with PAN and the 11 healthy controls were restudied on separate occasions. Following nasogastric intubation performed after an overnight fast, patients were studied during a 40-min basal period and for 60 min following SC injection of 0.06 mg/kg bethanechol chloride (Urecholine, Merck, Sharp and Dohme). This dose is effective in stimulating lower esophageal pressure (Roling et al. 1972) and does not produce any undesirable side-effects. Acid output was calculated in microequivalents per minute, fluid output in milliliters per minute, and gastric fractional emptying rates in percent per minute.

Two-factor (group of subjects and time) analysis of variance with repeated measurements (Kirk et al. 1968), the program LDU 040 (K. L. Dorn), and an IBM 370 computer (Division of Computer Research and Technology, National Institutes of Health, Bethesda, MD) were used to evaluate the statistical significance of differences observed for each function (e.g., acid output, fluid output, etc.)

Results

During fasting, gastric fractional emptying rate and acid output in 14 PAN patients were twofold less than that of healthy controls ($P < 0.05$), but their fluid output was not significantly altered. Intragastric volume was slightly but not significantly greater in patients with PAN (22 $\pm$ 4 ml) than in controls (16 $\pm$ 3 ml).

During pentagastrin infusion the fractional emptying rate tended to be less in 14 PAN patients than in controls, but the difference was not statistically significant. This supramaximal dose of pentagastrin produced significantly less acid ($P < 0.05$) and fluid output ($P < 0.01$) in 14 PAN patients than in controls. As a result, intragastric volume was significantly less in 14 PAN patients (78 $\pm$ 13 ml) than in controls (113 $\pm$ 6 ml; $P < 0.05$).

Following a water load, fractional emptying rate, acid output, and fluid output were significantly less ($P < 0.05$) in PAN patients than in controls, resulting in significantly greater intragastric volumes in 14 PAN patients than in controls ($P < 0.05$; Fig. 1).

One patient (K. H.) was not included with the other patients, since she had dramatically greater fractional emptying rates (19.2%/min during basal period, 24%/min during pentagastrin infusion, 35.5%/min following the load) and basal fluid output (2.3 ml/min) than the other 14 PAN patients and healthy controls.

Gastric Function in Primary Anorexia Nervosa 89

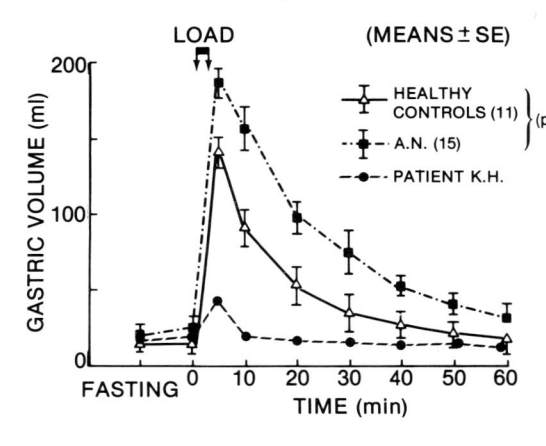

Fig. 1. Gastric volume following a 250-ml water load in 11 healthy controls, 14 PAN patients, and in one "different" PAN patient, K. H. Reproduced from (Dubois et al. 1979, with permission of the editor of Gastroenterology)

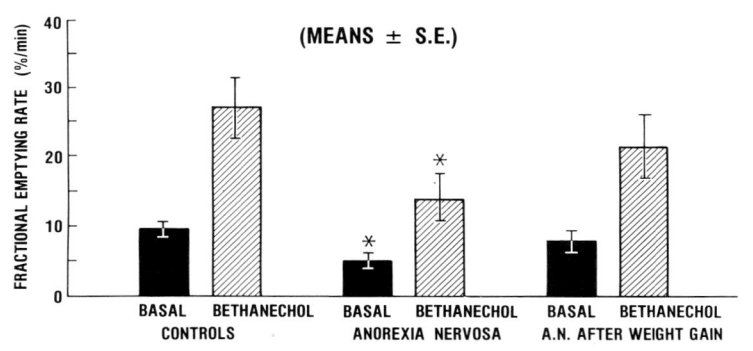

Fig. 2. Effect of bethanechol on gastric fractional emptying rate in 11 healthy controls and five patients with PAN before and after weight gain (*$P < 0.05$ compared with healthy controls during a basal period). (Reproduced from Dubois et al. 1981 with permission of the editor of Digestive Diseases and Sciences)

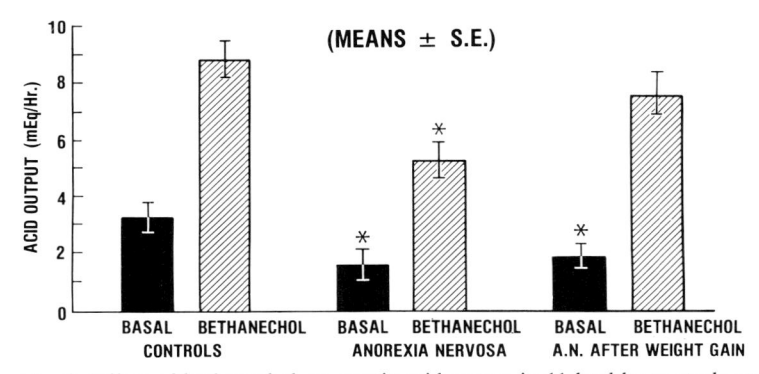

Fig. 3. Effect of bethanechol on gastric acid output in 11 healthy controls and five patients with primary anorexia nervosa (PAN) before and after weight gain (*$P < 0.05$ compared with healthy controls during basal or bethanechol periods). (Reproduced from Dubois et al. 1981 with permission of the editor of Digestive Diseases and Sciences)

Postpentagastrin and postload intragastric volumes were dramatically less than in controls and the other 14 PAN patients (Fig. 1). K. H. did not appear to be different from the others with respect to acid output or to degree of psychiatric symptomatology, particularly depression.

Six PAN patients were studied a second time following weight gain of 10−13 kg with a mean of 11 ± 3 kg. K. H. was not restudied, as she was lost to follow-up. These six patients showed no significant change in gastric acid output and a nonsignificant trend of fractional emptying rate to return toward control values. As a result, fractional emptying after weight gain was still significantly less than in controls following the load ($P < 0.05$) but not during the basal period.

Fractional gastric emptying rate and acid output were also less in PAN patients than in controls after bethanechol ($P < 0.05$; Figs. 2 and 3). However, bethanechol increased fractional emptying and acid output threefold in both controls and PAN patients. In both groups, the peak values were observed between 10 and 30 min after bethanechol for both gastric parameters and lasted 10−20 min. Thus, administration of bethanechol in PAN patients increased fractional emptying rates and acid output to a value similar to normal basal levels (Figs. 2 and 3). Following weight gain, all parameters were still slightly less in PAN patients than in controls, but the difference was statistically significant only for basal acid output (Figs. 2 and 3).

Discussion

In our studies, fractional gastric emptying rate was significantly less in 14 of 15 PAN patients than in controls during basal conditions and following a water load, but not during the infusion of maximal doses of pentagastrin. This decreased emptying in PAN patients could be related to their lower weight, as emptying is inversely correlated with body weight in healthy controls (Lavigne et al. 1978) and as obese subjects may have an abnormally rapid gastric emptying (Wright et al. 1983). Using barium meal X-ray examination, Scobie (1973) found duodenal dilatation in 10 of 20 PAN patients, but no delay in gastric emptying. Similarly, there was no significant alteration in basal gastric "motility" as determined using a highly compliant, continuously perfused catheter system with slow infusion rate (Silverstone and Russell 1967). However, these methods provide only qualitative and indirect information about gastric emptying, and are unreliable and probably unphysiologic. Recently, our results have been confirmed by two independent groups who found that gastric emptying of both liquids and solids was impaired (Saleh and Lebwohl 1980; Holt et al. 1981).

As a result of the decreased fractional emptying, postload intragastric volumes were greater in 14 of 15 PAN patients, than in controls (Fig. 1). Abnormally great intragastric volumes following a water load, as well as after solid meals (Holt et al. 1981; Saleh and Lebwohl 1980) may explain early satiety and acute gastric dilatation, as reported during refeeding of PAN patients (Evans 1968; Jennings and Klidjian 1974; Russell 1966; Scobie 1973), as well as the postprandial fullness described by most patients. However, intragastric administration of the water load produced discomfort and pain during the test only in the patient (K. H.) who had increased fractional emptying, whereas the other 14 did not report any symptoms following the load. This unique observation suggests that postprandial discomfort and pain found in some patients with PAN is not related to gastric distension, but could result rather from the rapid emptying of food into the intestines.

Basal and stimulated acid output were significantly less in PAN patients than in controls. This observation suggests that parietal cell mass in our PAN patients was less than in controls, and this could be related to the lesser weight of these patients. However, H^+ output had not returned to control values following an increase in body weight from 34 to 45 kg. Malnutrition found in most PAN patients could be responsible for the observed gastric hyposecretion, as suggested by the observation that pentagastrin-stimulated acid output was decreased in rats subjected to selective dietary protein deprivation (Thomasen et al. 1981).

After weight gain, fractional emptying rate tended to increase toward control values, but remained significantly lower than control following a water load. This incomplete recovery, at a time when most other somatic parameters such as hypotension and bradycardia (Gross et al. 1979) have returned to control values, suggests that gastric emptying is abnormally low in patients with PAN, even after weight gain. However, this hypothesis should be tested by restudying the patients after normal nutritional status has been maintained for longer periods of time.

Finally, our observations demonstrate that acute cholinergic stimulation can temporarily increase gastric acid output and fractional emptying rate in PAN patients. Previous studies had shown that the IV infusion of cholinergic agents stimulated basal acid output in Heidenhain pouch dogs (Gillespie and Grossman 1964) and in normal men (Vatn et al. 1975). In addition, these same agents could potentiate the effect of pentagastrin (Roland et al. 1975), and also of gastrin and histamine (Gillespie and Grossman 1964). Similarly, cholinergic stimulation was found to increase gastric emptying in functional gastric retention (Vasconez et al. 1970). Thus, the present studies suggest that treating the disordered gastric function in these subjects may accelerate their rate of weight gain.

The stimulation of both gastric emptying and acid output following administration of cholinergic agents suggests that the gastric smooth muscle and parietal cells remain responsive to cholinergic stimulation in PAN patients. However, since the stimulated gastric parameters are less in PAN patients than in controls, the gastric effector cells may be atrophied or the sensitivity of their cholinergic receptors may be decreased. In order to establish one or the other abnormality, one would have to construct complete dose-response curves to bethanechol, and this cannot be performed in these delicate patients.

Cholinergic therapy may be an effective adjuvant treatment of primary anorexia nervosa, since bethanechol increases gastric fractional emptying and acid output in those patients to levels similar to the ones observed in healthy controls. This effect could partly explain the improvement found in PAN patients given metoclopramide (Moldowski et al. 1977; Saleh and Lebwohl

1980), as this drug both stimulates the release of acetylcholine and inhibits dopamine receptors.

References

Dubois A, Van Eerdewegh P, Gardner JD (1977) Gastric emptying in Zollinger-Ellison syndrome. J Clin Invest 59: 255–263
Dubois A, Gross HA, Ebert MH, Castell DO (1979) Altered gastric emptying and secretion in primary anorexia nervosa. Gastroenterology 77: 319–323
Dubois A, Gross HA, Richter JE, Ebert MH (1981) Effect of bethanechol on gastric functions in primary anorexia nervosa. Dig Dis Sci 26: 598–600
Evans DS (1968) Acute dilation and spontaneous rupture of the stomach. Br J Surg 55: 940–942
Feighner JP, Robins E, Guze S (1972) Diagnostic criteria for use in psychiatric research. Arch Gen Psychiatry 26: 57–63
Gillespie IE, Grossman MI (1964) Potentiation between urecholine and gastrin extract and between urecholine and histamine in the stimulations of Heidenhain pouches. Gut 5: 71–76
Gross HA, Lake CR, Ebert MH, Ziegler M, Kopin IJ (1979) Catecholamine metabolism in primary anorexia nervosa. J Clin Endocrinol Metab 49: 805–809
Holt S, Ford MJ, Grant S (1981) Abnormal gastric emptying in primary anorexia nervosa. Br J Psychiatry 139: 550–552
Jennings KP, Klidjian AM (1974) Acute gastric dilatation in anorexia nervosa. Br Med J 2: 477–478
Keys A, Brozek J, Henschel A (1950) The biology of human starvation. University of Minnesota Press, Minneapolis, pp 587–600
Kirk RE (1968) Experimental design: procedures for the behavioral sciences. Brooks Cole, Monterey, pp 110–114
Lavigne ME, Wiley ZD, Meyer JH (1978) Gastric emptying rates of solid food in relation to body size. Gastroenterology 74: 1258–1260
Markowski B (1947) Acute dilation of the stomach. Br Med J 2: 128–130
Moldofsky H, Jeuniewic N, Garfinkel PE (1977) Preliminary report on metoclopramide in anorexia nervosa. In: Vigersky RA (ed) Anorexia nervosa. Raven, New York, pp 373–375
Roland M, Berstad A, Liavag I (1975) Effect of carbacholine and urecholine on pentagastrin-stimulated gastric secretion in healthy subjects. Scand J Gastroenterol 10: 357–362
Roling GT, Farrell RL, Castell DO (1972) Cholinergic response of the lower esophageal sphincter. Am J Physiol 222: 967–972
Russell GFM (1966) Acute dilatation of the stomach in anorexia nervosa. Br J Psychiatry 112: 203–207
Saleh JW, Lebwohl P (1980) Metoclopramide-induced gastric emptying in patients with anorexia nervosa. Am J Gastroenterol 74: 127–132
Scobie BA (1973) Acute gastric dilatation and duodenal ileus in anorexia nervosa. Med J Aust 2: 932–934
Silverstone JT, Russell GFM (1967) Gastric "hunger" contractions in anorexia nervosa. Br J Psychiatry 113: 257–263
Thomason H, Burke V, Gracey M (1981) Impaired gastric function in experimental malnutrition. Am J Clin Nutr 34: 1278–1280
Vasconez LO, Adams JT, Woodward ER (1970) Treatment of reluctant postvagotomy stoma with bethanechol. Arch Surg 100: 693–694
Vatn MH, Schrumpf E, Myren J (1975) The effect of carbachol and pentagastrin on the gastric secretion of acid, pepsin and intrinsic factor (IF) in man. Scand J Gastroenterol 10: 55–58
Wright RA, Krinsky S, Fleeman C, Trujillo J (1983) Gastric emptying and obesity. Gastroenterology 84: 747–751

Psychophysiological Indices of the Feeding Response in Anorexia Nervosa Patients

R. Hölzl and S. Lautenbacher

Introduction

The Role of Afferent Signals from the Gastrointestinal Tract in Food Intake Regulation

The question as to whether and how afferent signals from the gastrointestinal tract (GIT), and from the stomach in particular, regulate food intake has been a classic issue at least since Cannon and Washburn (1912). Despite later refutations of their concept of "hunger contractions" as the direct stimulus to eat, and some controversy between "centralists" and "peripheralists" in the animal literature, the role of feedback from the GIT in food intake regulation is now generally accepted (Konturek and Rösch 1976; Booth 1978). Neglecting effects of "conditioned hunger" and "appetites" for the moment, it seems that initiation of eating is governed mainly by hypothalamic centers according to humoral factors. Termination of meals, however, is under the control of a "fast feedback loop", in which afferent information from the GIT about gastric filling and nutritional composition of food ingested serves as control variable in conjunction with olfactory and gustatory stimuli arising during oral stages of ingestion. Contents absorbed from the intestines and blood concentrations of nutritional substances provide second- and third-stage feedback signals with longer time constants. They mainly determine the length of interdigestive intervals and not termination of intake.

Local control of stomach emptying and secretion by nutritional contents and other variables contributes to meal patterning and hunger regulation by influencing digestion times. Fatty food, for instance, retards stomach emptying. It enters the small intestine later and at a slower rate. This causes prolonged feedback from the small intestine and in turn a longer feeding pause. In this way retardation of emptying reduces food intake.

Although specific hunger contractions have not been found in more recent studies, more frequent and stronger stomach contractions towards the end of interdigestive intervals have consistently been observed (Konturek and Rösch 1976). It seems probably that this increased GI activity serves as an afferent signal which enhances probability of meal initiation or appetitive behavior (Hollis 1982). In this form, the core of the earlier assumption that GI motility controls food intake by determining not only termination of but also initiation of meals might be worth retaining. A self-stimulation experiment by Ball (1974) at

---

1 Max-Planck-Institut für Psychiatrie, Kraepelinstraße 10, D-8000 München 40

least clearly shows that afferent signals from the stomach directly influence the activity of the "hunger center" in the lateral hypothalamus: after the gastric branch of the vagus had been severed in rats trained to press a lever and reinforced by electrical stimulation of the lateral hypothalamus, self-stimulation thresholds were raised by magnitudes. This suggests conjunctive operation of hypothalamic and gastrointestinal activity in control of uptake. The particular interpretations of this and related findings are debated, however (Berger 1977; Kupfermann 1981).

In a brilliant integration of ethological and conditioning literature, Hollis (1982) drew attention to a further mechanism by which GI variables, and motility in particular, interact with central factors in eating control. The author shows convincingly that GI activity is an integrated part of appetitive behavior, and that this integration takes place by Pavlovian conditioning of this activity to food-related stimuli and/or species-specific behavior elicited by them or conditioned to them. This learning process also determines the appetitive or reinforcing value of different kinds of food. According to Hollis (1982) and the literature cited by her, the attractiveness of food depends not only on the deprivation state of the animal and stimuli directly correlated with taste, smell, etc. of food itself, but also on various stages of appetitive and consummatory behavior, including GI responses classically conditioned to environmental, internal, or response-produced stimuli.

Gastrointestinal Activity and Eating Disorders

Given the experimental evidence cited above there must be several ways in which altered GI activity could contribute to disturbed eating control in anorectic and obese patients. Stunkard and his co-workers were among the first to discuss the possible role of diminished sensitivity to afferent stimuli from the GIT in eating disorders. In an attempt to corroborate Schachter's findings that obese patients depended more on exteroceptive than on interoceptive stimuli in regulating their food intake, however, they found no evidence for diminished interoceptive sensitivity to spontaneous stomach contractions in obese compared with control subjects, despite a stronger response bias as measured by their signal detection method (Stunkard and Fox 1974). The adequacy of Stunkard's approach was later debated because of the irritating nasopharyngeal tube with which this group tried to measure stomach contractions. Tube-related stimuli may well mask more subtle interoceptive signals. In a preliminary study in our laboratory (Striegel 1978) using noninvasive surface gastrography Stunkard's results were replicated, however. Obese *and* anorectic patients were compared with controls. Differences from controls were found only with subjective hunger ratings and motility base rates and not in interoception indices. The methods and paradigms used were primitive, and the issue is by no means settled. On the contrary, work by Bruch (1973) and Garfinkel et al. (1978) (Garfinkel and Garner, this volume) still supports the notion of altered interoception in eating disorders, but the precise mechanism remains to be investigated.

The measurements of fasting activity and its perception may not be relevant to the understanding of dysregulation of food intake at all (Crisp 1965). At present, therefore, investigation of the *feeding response* in anorectic patients at the physiological, behavioral, and subjective levels seems more promising. Little is known about the range of variations in GI and other autonomic changes to a meal requirement in these patients, how these changes are reflected on the subjective level, and whether this pattern undergoes systematic alterations during weight gain procedures. Only the background of a more thorough knowledge of these responses would make it possible to interpret perceptional differences. Differences in base rates or the contribution of conditioned feeding responses could then be separated from alterations in interoception proper.

Dubois et al. (1979) seem to have been the first to measure gastric emptying rates in anorectic patients with sufficient precision. Their method ("fractional emptying") simultaneously measures gastric secretion and takes its contribution to stomach contents into account. The authors found retarded emptying and changes in acid output in their patients. On the assumption that this was a significant factor in maintenance of the disorder if not in its etiology, Saleh and Lebwohl (1980) attempted acceleration of gastric transit in anorectics by oral administration of metoclopramide and found apparent improvement in some of their patients. Although Dubois et al. (1981) could not generally support these results in a similar study using bethanechol, the importance of gastric function in maintaining dieting seems to be established (cf. Dubois et al., this volume).

Whether the results on emptying also extend to motility records remains to be tested. An earlier investigation by Crisp (1965) revealed no differences in motility compared with controls, but the author used an intragastric tube in fasting subjects. Nothing can be said about postprandial motility and possible inhibitory side-effects of the tube. They may prevent detection of differences from controls by reducing base rates of contractions in both groups to near zero level, thus giving rise to a "floor effect". In fact, this is exactly what was found in the studies cited. Therefore it would be advisable to investigate this question by noninvasive techniques such as surface gastrography. Reliable and valid techniques for certain periodic motility components are now available.

The present chapter reports preliminary data on the measurement of the feeding response in anorectic patients by means of a variant of surface gastrography reported elsewhere (Hölzl 1983; Müller et al. 1983). The method produces reliable and valid indices of specified periodic components of electromotor activity of the stomach without the adverse effects of intragastric pressure probes. In addition it is not negatively affected by nonfluid gastric contents. It therefore allows undisturbed recording before, during, and after a realistic test meal, in contrast to Crisp's fasting records or Dubois' method, for which a test drink of water was necessary.

Changes in motility parameters after a test meal have been described by several researchers. Those were related to emptying in healthy subjects and certain clinical groups other than anorectics. Usually a typical biphasic change in frequency of periodic contractions (Fig. 4) and a postprandial increase in amplitude is seen (e.g., Kohatsu 1970; Konturek and Rösch 1976; Smout 1980). From Dubois' results on emptying one might conclude that the time course of motility changes after a meal will also be delayed. This would be related to an increased and prolonged subjective feeling of "fullness" if visceral perception remained undisturbed. Frequency and amplitude of postprandial contractions, however, are not correlated with emptying rate in a simple linear fashion. Pyloric resistance, regularity of pressure waves, and antropyloric coordination also influence emptying. Sometimes "functional stenosis" has been connected with retarded emptying in anorexia nervosa. Therefore more complicated alterations of motility patterns might also exist in these patients.

If retardation of the feeding response as measured by motility records was mainly an adaptation to chronic dieting but had no primary etiological significance for anorexia the delay would systematically decrease with weight gain during behavioral treatment. On the other hand, if alterations of the feeding response played a role in initial pathogenesis it would be distinguishable from simple adaptations of the GIT to malnutrition. And lastly, the degree of recovery of a normal feeding response through treatment would be inversely related to risk of anorectic relapses. If these relationships could be shown to hold for the feeding response indices chosen, they could be used not only for diagnostic and research purposes but also to guide therapy and reduce relapses by timely booster treatments. The practical implications of this type of psychophysiological assessment are obvious.

To distinguish direct GI components of the feeding response from autonomic correlates of emotional responses to the meal requirement which in these patients presumably is aversive, recording of other autonomic variables such as heart rate and ratings of task-related subjective states is mandatory. The recording of cardiovascular components of the feeding response has further advantages. Evoked heart rate responses to a meal were earlier found to be correlated with its motivational significance. Eisman (1966), for instance, showed that only thirsty rats showed phasic heart rate increases to the presentation of water, while rats with free access to water did not change their heart rates. Similar results were reported in a comparison of hungry and satiated human subjects. Evaluation of evoked heart rate responses to a test meal in anorectics might thus give important additional and "objective" information on the motivational state. Furthermore, a particular cardiorespiratory parameter, the degree of cardiac-respiratory coupling, has been shown to be a useful and selective index of parasympathetic heart control or "vagotonus" (Hölzl et al. 1983). This index will be used as a measure of autonomic activation by the feeding test in the patient.

Method

To test some of the hypotheses listed above, and the question of retarded feeding responses in anorexia patients in particular, so far 14 patients taking part in the MPIP Anorexia Project, 12 females and two males, and 10 healthy controls have been studied. For six patients and nine controls an ininterrupted series of three test sessions at 2-week intervals is available at present. Control subjects belonged to two groups: K1 consisted of five nonanorectic persons at least 15% below normal in weight (4 females, 1 male). This group was chosen to control for possible effects of low body weight and corresponding changes in body build on measures of motility. Such relations had been found in previous studies (Skambraks 1981). K2 contained five healthy subjects volunteering in a weight reduction program (Pirke et al., to be published). K2 was included in the experimental design to compare anorectic feeding responses with feeding responses in dieting subjects according to the rationale discussed in the *Introduction*. Weight ranges at the start of the study were 48.9%−77.0% of normal weight for the patient group, 70.0%−85.9% for K1, and 80%−92.6% for K2. For inclusion in group K2 subjects were required to have body weight within 5% of their ideal

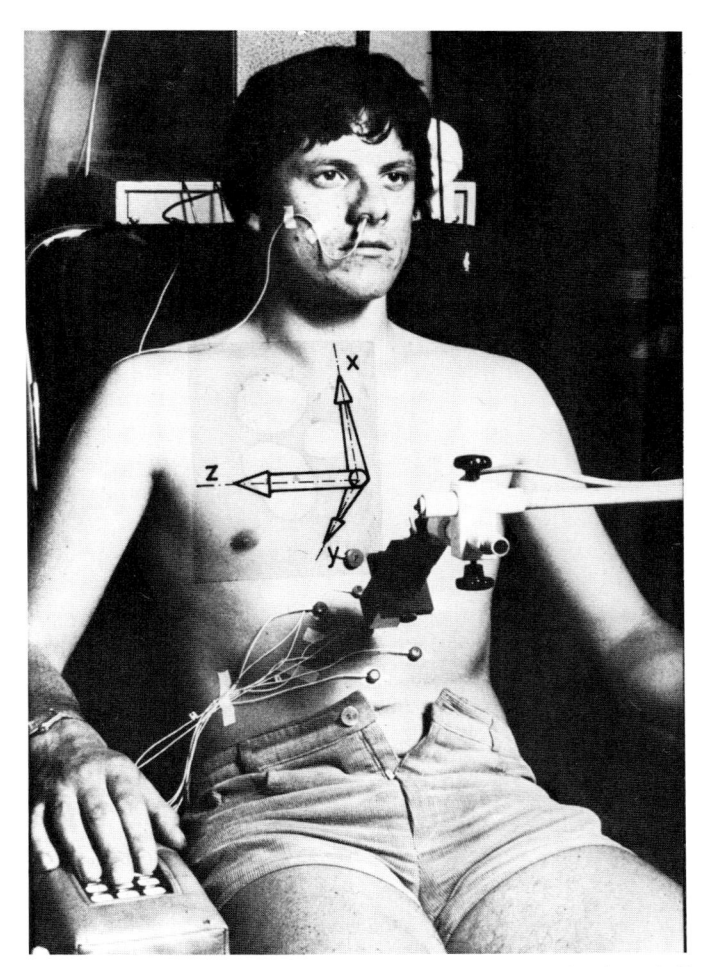

Fig. 1. Experimental subject for conjoint gastrography in situ. A Develco three-axes fluxgate magnetometer is placed over the epigastric region, orthogonal to the abdominal plane. Axes are indicated by *arrows*. Beneath the magnetometer head the hexagonal electrode configuration for bipolar electrogastrograms can be seen. In addition ground and ECG electrodes are placed on the sternum but only partially shown. The probe in the nose contains a thermistor to measure respiration. Technical details are described in Müller et al. (1983), from which this illustration is reproduced with kind permission of Plenum, New York

weight (= normal weight −15%) at the initial sessions. Subjects in this group were paid to reduce this initial weight by 10% within 6 weeks. The average ages of groups were not matched exactly, patients being somewhat younger than controls (21 vs 25); ranges were comparable.

All subjects had fasted overnight since 10 p.m. the previous evening before testing. Sessions started invariably at 7.30 a.m. They consisted of six epochs each 10 min in length and an additional preliminary adaptation period of the same duration. The six recording periods also served as analysis epochs in biosignal evaluation (see below). At the beginning of the second recording period a test meal of 250 g yoghurt was given, which was followed by 40 min of postprandial recording. Subjective reactions to the meal were evaluated by five-point ratings of "fullness", "hunger", and general "tension" at the end of each 10-min interval.

Gastric motility was measured by "conjoint spectral gastrography" (CSG) (Müller et al. 1983). This method consists of simultaneous multichannel recordings of six channels of bipolar and one channel of unipolar surface electrogastrograms (EGGs) and of three channels of magnetogastrograms (MGGs). The latter are obtained with a three-axes fluxgate magnetometer. It senses stomach contractions via magnetic field changes induced by a small teflon-coated magnet which the subject swallows before the recording. EGGs and MGGs are jointly subjected to Fourier analysis. The

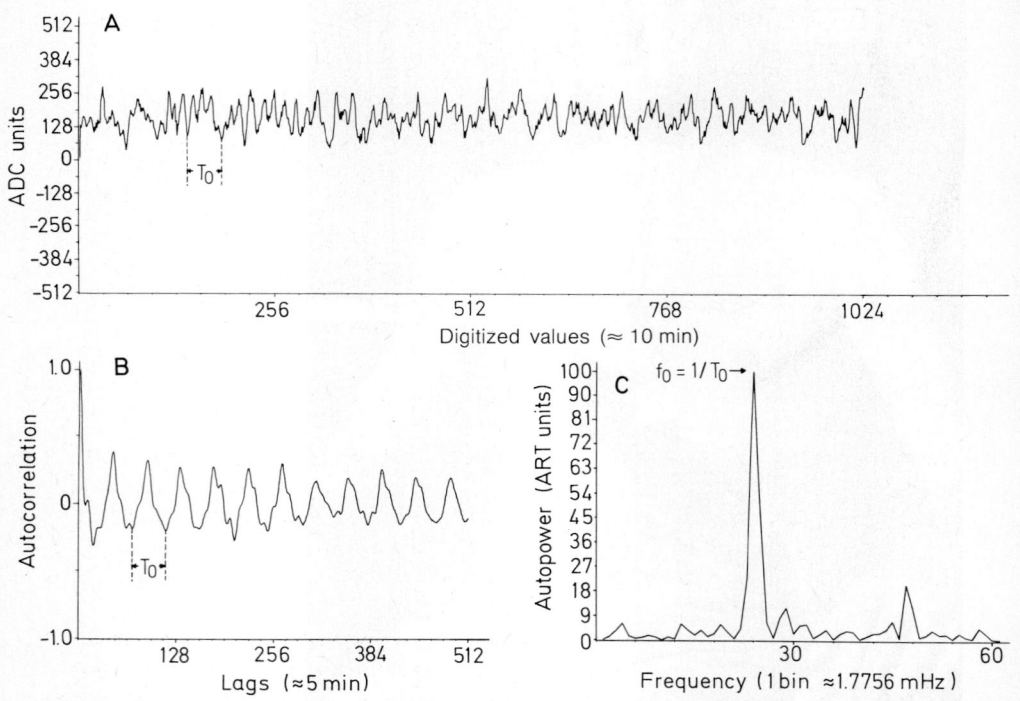

Fig. 2A–C. Sample of surface electrogastrogram, its autocorrelation function, and corresponding power density spectrum. **A** Original recording with respiratory artefact and noise as digitized by the A-D converter (512 ADC units = 500 µV). Sampling intervals is 550 ms; recording period = 1,024 samples or approx. 10 min; T_0 = slow wave period = approx. 20 s. **B** Autocorrelation function (ACF) of the data in record **A** with lags extending up to 5 min. The ACF accentuates the periodic slow wave component and reduces noise. Again T_0 = slow wave period. **C** The auto-power spectrum of record **A** shows a pronounced peak at the frequency of the slow wave component, $f_0 = 1/T_0 =$ 2.6 cpm, a smaller peak at the first "harmonic" ($2f_0$), and only minor noise energy at other frequencies. The absolute or relative height of the slow wave peak gives an estimate of slow wave intensity. See also Müller et al. (1983) for technical details

analysis produces reliable and valid measures of frequency and amplitude of specified periodic components of gastric smooth muscle activity for each 10-min analysis epoch. Particularly two frequency bands, the basal gastric rhythm (BGR) at 3 cpm and the ultraslow gastric rhythm (UGR) at 1 cpm, are evaluated in this way. Changes of frequency and amplitude of these rhythms during and after meals are interpreted as the gastrographic feeding response.

A detailed description of the method, its rationale, and relevant validation data have been published elsewhere (Müller et al. 1983). Therefore a short illustration of the method will suffice here. Figure 1 shows an experimental subject with the positions of abdominal electrodes for hexagonal bipolar EGG recordings. The magnetometer is positioned orthogonal to the abdominal plane just over the middle electrode. Magnetic axes are indicated by arrows. The thermistor probe in the nose senses respiratory air flow. Sternal electrodes are used to record the ECG from which heart rate is calculated. Figure 2 shows a sample of a typical gastrographic recording together with its autocorrelation function and corresponding power spectrum. The pronounced peak in the spectrum corresponds to the BGR at 3 cpm, which is the clear periodicity in the raw record. The periodicity is even more pronounced in the autocorrelation function. The frequency and amplitude of these peaks are evaluated for each 10-min epoch and enter analysis of the feeding response.

In addition to gastrographic changes, cardiac and respiratory activity were measured during feeding test. Signal parameters entering statistical analysis for each 10-min epoch were average heart rate, respiratory frequency and amplitude, and Porges' index of cardiorespiratory coupling, i.e., the "weighted coherence" (C_w) between heart rate and respiratory air flow (Hölzl et al. 1983; Porges et al. 1980).

Psychophysiological variables and subjective ratings were analysed as individual and group reaction patterns over number of epoch (1–6) as the independent variable. Raw values and absolute and percentage cumulative feeding responses were considered. The relative or percentage cumulative feeding response is of particular value in comparing the time course of feeding responses in anorectics and controls without confusion caused by differences in absolute reaction levels. Group data analyses are therefore mostly based on relative cumulative records (further explanation below).

Results

Individual Response Patterns

Because little is known about the variation range of postprandial motility changes as measured by gastrography, individual response patterns were studied on a descriptive level first. Typical examples of this descriptive analysis of a) amplitude (or root-power) spectra changes during the feeding test; b) BGR frequency patterns; c) BGR amplitude responses; d) cardiorespiratory effects; and e) subjective ratings are given below.

Figure 3 illustrates the type of change in gastrographic spectra to be observed after a test meal of yoghurt in an anorectic patient and a comparable control subject more than 20% below normal body weight. Each spectrum represents the periodic content of 10 min of gastrographic recordings, and has characteristic intensity maxima at the frequency of the BGR component (approx. 3 cpm). Their height was normalized to make the positions of relative maxima in time more obvious. There are also pronounced peaks in the UGR range (1 cpm, to the left of the BGR peaks). They will not be analysed here. In the present example a clear delay of the BGR maximum by two epochs or about 20 min in the anorectic patient compared with the control was found. Systematic group data on this delay will be presented below.

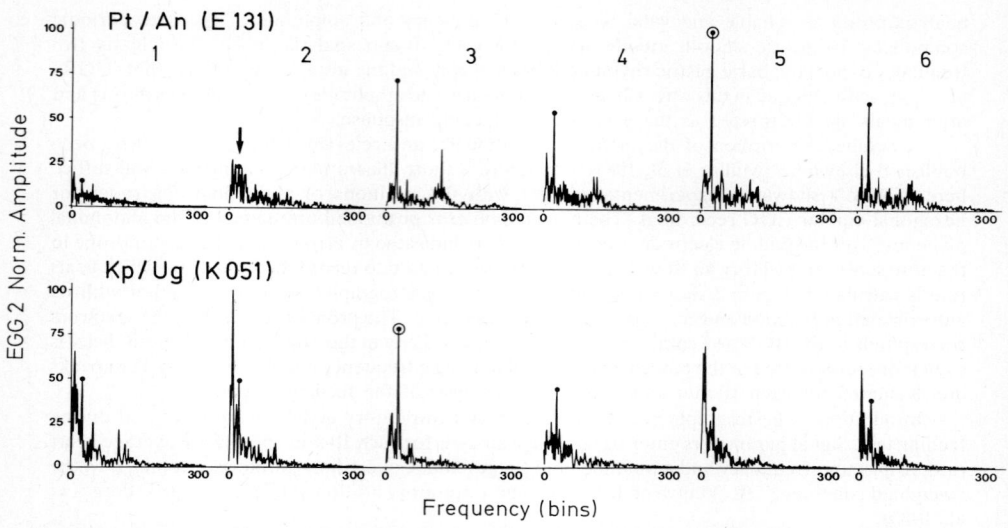

Fig. 3. Gastrographic "feeding response". Amplitude spectra of surface gastrograms show systematic changes in subsequent 10-min epochs of measurement during the feeding test. Spectrum 1 corresponds to preprandial, spectra 3 through 6 to postprandial periods. The test meal is ingested at the beginning of the second 10-min period. Examples show the feeding responses in session 1 of a control subject from group K1 *(lower half, Kp/Ug)* and a patient *(upper half, Pt/An)*. Amplitude peaks are normalized relative to the maximal peak within the individual feeding test. Frequencies are displayed as bins: 1 bin = 1.78 mHz. BGR components are marked with *dots* (●), maximum BGR peaks with *circles* (○). Current body weights were at 61% of normal for the patient and 70% for the nonanorectic control. The maximum postprandial rise of BGR amplitude is delayed in the patient by two epochs or approx. 20 min despite the relatively small difference in body weight

While amplitude changes of certain gastrographic rhythms can easily be seen in this type of data representation, frequency patterns are more difficult to detect. They are displayed separately in Fig. 4. Systematic biphasic changes of BGR frequencies are observed consistently by way of EGGs and of MGGs exemplifying the reliable detection of significant periodic components by CSG methods. The typical drop in BGR frequency during the meal and the subsequent rise over baseline shown in Fig. 4 is in good agreement with gastroenterological literature on direct recordings from implanted electrodes (Kohatsu 1970; Smout 1980). However, this is more difficult to show in anorectic subjects, because the detection logic of CSG depends on MGGs, which are frequently noisy. At present reproducible data on postprandial frequency changes of gastric rhythms are therefore difficult to obtain. Refinement of signal analysis will probably solve this problem. At this stage more reliable differences between patients and controls are seen in the postprandial BGR *amplitudes*. Figure 5 illustrates a biphasic pattern of BGR amplitude changes during feeding test in the control subject. The feeding response of the anorectic patient is different. It shows a sluggish onset, prolonged response, and late maxima in postprandial BGR amplitudes. In addition, the drop in amplitude during eating

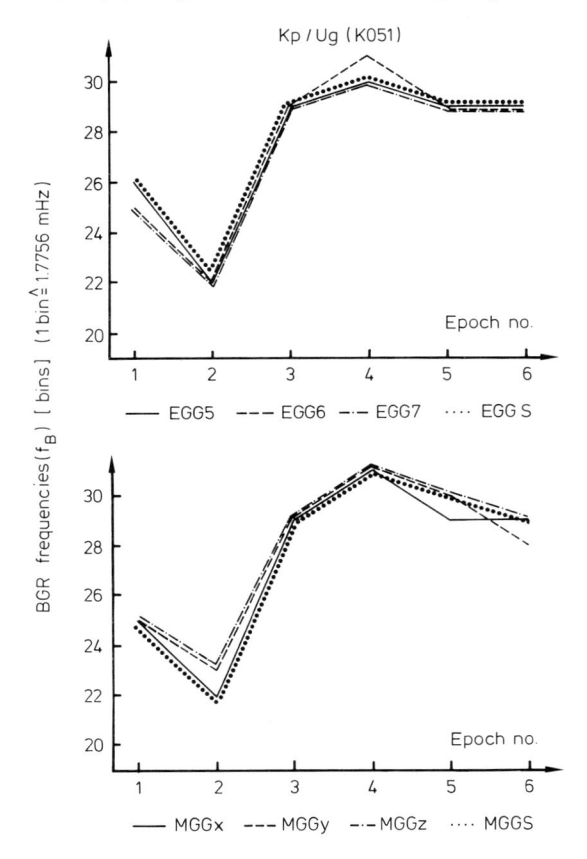

Fig. 4. Frequency changes of BGR components during feeding test. Data are from the same control subject as in Fig. 3. BGR frequencies of two bipolar EGGs (EGG 5 and 6) from the hexagon in Fig. 1, one unipolar EGG (EGG 7, leg reference), and a summated EGG (see Müller et al. 1983) are plotted in the *upper half* of the figure. The *lower half* shows BGR frequencies of MGGs. x, y, and z are the three axes in space, as explained in Fig. 1, MGGS is a summated MGG score

is not seen in the patient. But interindividual variability is still high, and other data representations will be used in later sections.

A comparison of subjective ratings with the gastrographic responses shows that common expectations about psychological covariates of delayed feeding responses may be wrong. As Fig. 6 demonstrates, subjective ratings of fullness are low throughout the session in the anorectic patient, whereas the control reports fullness after relatively rapid ingestion of 250 g yoghurt within first 3 min of the second epoch. This is really the opposite of what would intuitively be expected and of what gastrographic feeding responses would suggest. Hunger ratings shed some light on this paradox perhaps: the anorectic patient shows a surprisingly high hunger rating at the beginning of the session, which goes down after the test meal. This is not the case in the control subject, who shows constant low hunger throughout the session. A possible explanation would suggest that the patient's rating reflects what she considers socially desirable rather than her actual hunger or fullness. But more systematic data on larger groups are needed to corroborate these anecdotal findings.

The difference in tension ratings between patient and control is also interesting (Fig. 6). The control subject reports moderate activation, which only

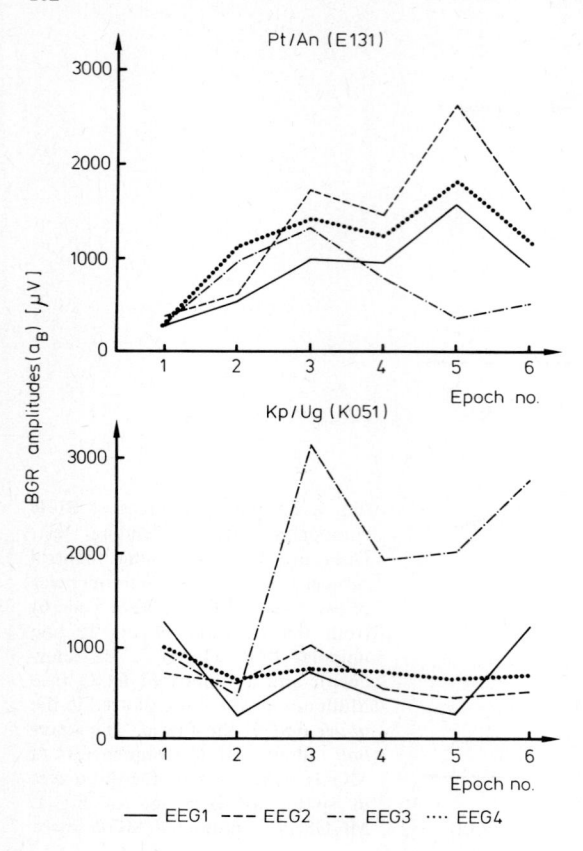

Fig. 5. BGR amplitude changes during feeding test. BGR peak heights from the same subjects as in Fig. 3 are shown for consecutive 10-min periods. Four bipolar EGGs are shown. Biphasic amplitude responses are observed in the control subject. The anorectic patient exhibits more or less monotonic increases in BGR amplitudes with sluggish onset and no inhibition during the meal

increases briefly during eating requirements. In contrast, the anorectic patient is tenser initially and only calms down later. Cardiac-respiratory responses parallel this pattern (Fig. 7). Respiratory frequency in the control subject shows only a phasic drop during the eating period. The same parameter drops monotonically in the patient. Average heart rate and cardiac-respiratory coupling, i.e., "vagotonus" measured by the C_w value, change phasically in both subjects. Despite subjective tenseness ratings suggesting no particular activation during the eating period a specific emotional effect of the eating requirement seems to be uncovered by the much larger drop in C_w in the patient. If this can be reproduced on a larger data base evaluation of parasympathetic heart control via C_w estimation may prove a valuable tool in elaborating the autonomic response pattern of anorectic patients to eating situations. Estimation of respiratory frequency in the eating period is unreliable, however, so that some precautions have to be taken in the future.

The illustrations of individual response patterns in the feeding test presented so far show several features worth studying in larger numbers and over longer intervals. Instability of gastrographic feeding responses over subjects, however,

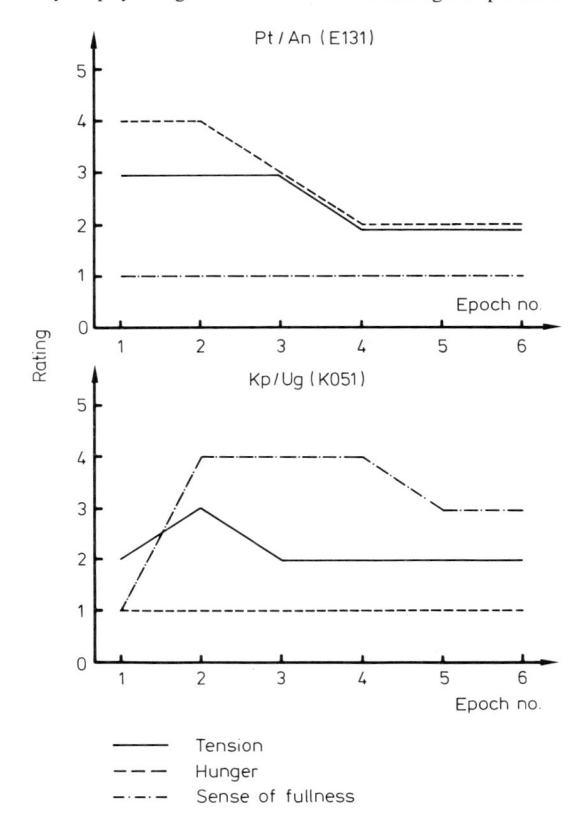

Fig. 6. Subjective ratings of hunger, tension, and fullness during the feeding test. Data from the same subjects as in previous figures

demands further developments in defining suitable indices of the feeding responses. One source of interindividual variation is large differences in initial values of physiologic variables which confound differences in response size. In group analyses *relative cumulative response graphs* are used below to account for initial level differences in physiological variables and thereby accentuate differences in the time course of the feeding response rather than absolute differences. For this purpose the cumulative values of physiologic parameters such as BGR amplitudes in each 10-min period are expressed as percentages of the final cumulative level of responses. Figure 8 exemplifies this kind of data representation for the patient and the control subject whose individual autonomic and subjective responses patterns were described above. The electrogastrographic curve shows changes in the electrical activity and the magnetogastrographic curve shows the contractile activity of gastric smooth muscles. The characteristic difference in the time course of the feeding response measured by cumulative BGR amplitudes is much enhanced, and the expected delay in gastrographic response to the meal is clearly demonstrated. Whether this holds for the group data is investigated in the next section.

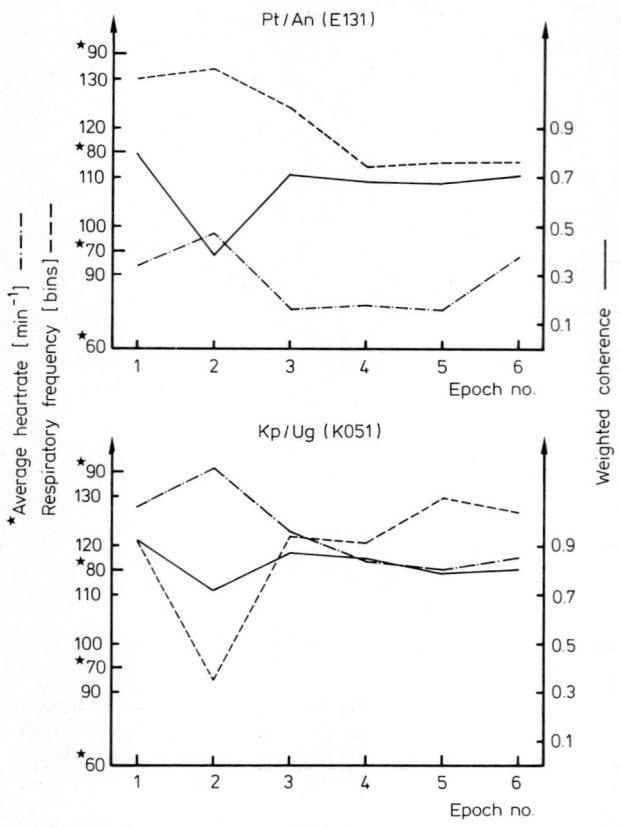

Fig. 7. Averaged heart rates, respiratory frequency, and cardiac-respiratory coupling during feeding test. Data stem from the same subjects as in previous figures. Analysis periods are 10 min as before. *Bold ordinates* on the *left* refer to heart rate, *fainter ordinates* to respiration frequency. Ordinates on the *right* indicate the dimensionless measure of cardiac-respiratory coupling, i.e., weighted coherence values, C_w, varying from 0 to 1 (explanation in text)

Group Differences in BGR Amplitude Changes to the Test Meal

Comparison of the *averaged cumulative response graphs* in the first session of anorectic patients and K1 and K2 controls combined reveals characteristic initial differences in the speed of feeding responses (Fig. 9). Cumulative response graphs express this in differences of slopes: cumulative feeding curves of control groups rise much faster and with negative acceleration, whereas the curves for anorectic patients start more slowly and have positive acceleration. As expected, the two control groups do not differ in the first session, because K2 subjects have only just started dieting at this point. K1 and K2 were therefore pooled for this graph and statistical testing. Percentages of total cumulative feeding response reached at the end of 3rd and 4th 10-min periods (1st and 2nd postprandial period) are used in the statistical testing as indices of the "time constants" of

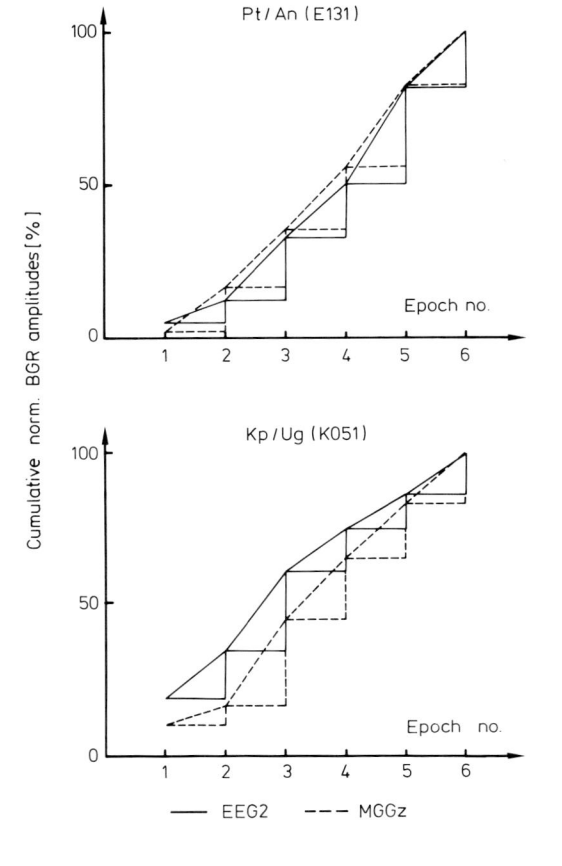

Fig. 8. Normalized cumulative feeding response. To enhance differences in the time course of BGR amplitude during the feeding test normalized cumulative plots are used. One EGG and one MGG are shown for illustration. Data stem from the same subjects as in previous figures. BGR amplitude values are cumulated over consecutive measurement periods and expressed as percentages of the final value of the cumulative series. Data are drawn in as polygon as well as step functions. The latter facilitates comparison of individual contributions of consecutive 10-min intervals

feeding responses. Group differences in these parameters are significant at the 5% level for some gastrographic channels but not for all (t-tests; see Table 1).

In addition to these initial differences in gastrointestinal effects of the test meal, the pattern of differences shows systematic changes over sessions which parallel weight gain in anorectics, weigth loss in dieting subjects, and constant weight in K1 subjects. Retardation of anorectic feeding responses gradually diminishes with weight gains, which range from −2.0% to 8.1% of normal weight from the 1st to the 2nd, and from 1.0% to 24.4% from the 2nd to the 3rd session. The simple explanation of the retardation as an adaptation to prolonged dieting is made improbable, however, by the feeding pattern in K2, i.e., nonanorectic subjects voluntarily reducing weight for a money fee. Weight reductions in this group range from −3.7% to −7.0% of normal weight from the 1st to the 2nd, and from −2.0% to −3.8% from the 2nd to the 3rd session. Instead of retardation of gastrographic feeding response, cumulative graphs of 3rd sessions after at least 6 weeks of low calory intake (see Pirke et al., to be published) indicate accelerated postprandial reaction patterns (Fig. 10). Because of the small numbers in this study differences do not reach conventional

Cumulative normed BGR amplitudes: EGG2, Session 1 (AS)

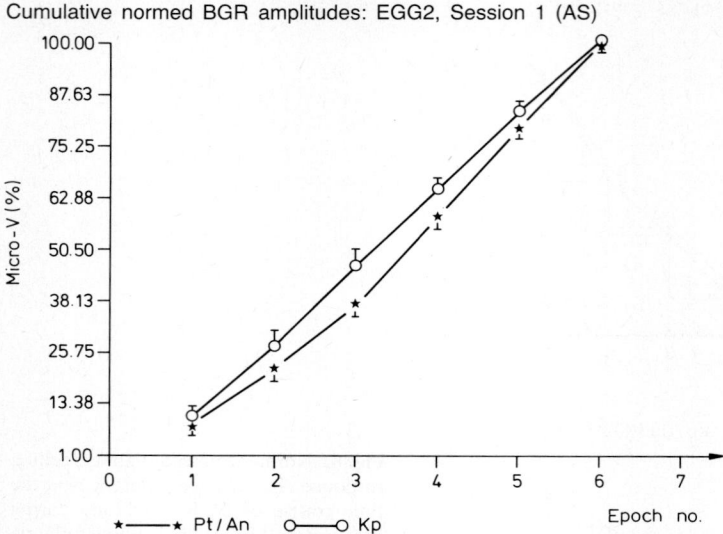

Fig. 9. Average group cumulative feeding responses in first session for anorectic patients and controls. All patients (PT/AN, $n = 14$) and control subjects from K1 and K2 (KP, $n = 10$) are included. Differences in the 3rd and 4th intervals are significant (t, 5%). *Vertical bars* indicate standard error of mean. See text and previous figures

Table 1. Group means, standard deviations, and significance testing of relative index of feeding response speed during first session (cumulative percentage of BGR amplitude, EGG2; see text)

Measurement period	Patients ($n = 14$)	K1: Thin controls ($n = 5$)	K2: Diet controls ($n = 5$)	K1 and K2 ($n = 10$)
1	7.86 (4.24)	12.80 (5.54)	6.20 (1.92)	9.50 (5.23)
2	21.36 (8.60)	34.20 (8.93)	19.60 (6.69)	26.90 (10.70)
3	37.00 (8.17)	50.80** (9.04)	41.60 (11.22)	46.20* (10.76)
4	58.21 (7.80)	65.60+ (8.08)	63.60 (6.88)	64.60* (7.15)
5	79.07 (4.95)	84.40 (4.77)	81.80 (4.32)	83.10 (4.51)

Statistically significant differences in control subjects as against anorectic patients: + $P < 0.10$; * $P < 0.05$; ** $P < 0.01$ (t-test)

significance levels in the 3rd sessions, and further experiments will be necessary. Group differences are better accentuated in the time constant indices defined above. But group differences are significant at the 5% level only for the first session (t-tests; see Table 2 and Fig. 11).

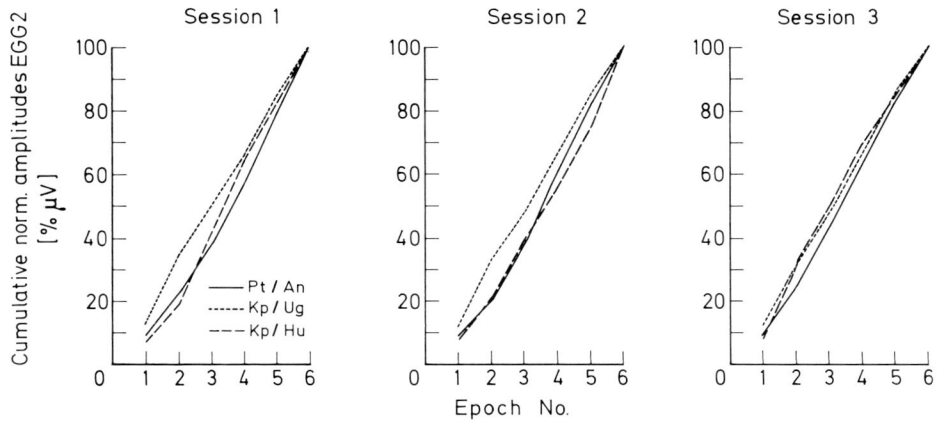

Fig. 10. Changes in cumulative group responses over three sessions. *PT/AN*, patients; *KP/UG*, K1; *KP/HU*, K2. See text and previous figures for further explanation

Table 2. Group means, standard deviations, and significance testing of relative index of feeding response speed over three sessions (cumulative percentage of BGR amplitude, EGG2; see text)

Session	Patients ($n = 6$)	K1: Thin controls ($n = 5$)	K2: Diet controls ($n = 4$)	K1 and K2 ($n = 9$)
(a) 1st Postprandial interval				
1	38.17 (6.52)	50.80* (9.04)	42.00 (12.91)	46.89 (11.17)
2	38.17 (14.08)	47.40 (12.66)	39.00 (7.75)	43.67 (11.06)
3	42.83 (10.98)	48.00 (14.61)	48.75 (15.33)	48.33 (48.33)
(b) 2nd Postprandial interval				
1	57.33 (6.02)	65.60** (8.08)	63.75 (7.93)	64.78 (7.56)
2	61.00 (11.24)	66.00 (10.70)	55.75 (9.03)	61.44 (10.82)
3	63.33 (7.37)	66.20 (9.26)	69.00 (9.93)	67.44 (9.06)

Statistically significant differences in control subjects as against anorectic patients: * $P < 0.05$; ** $P < 0.10$ (*t*-test)

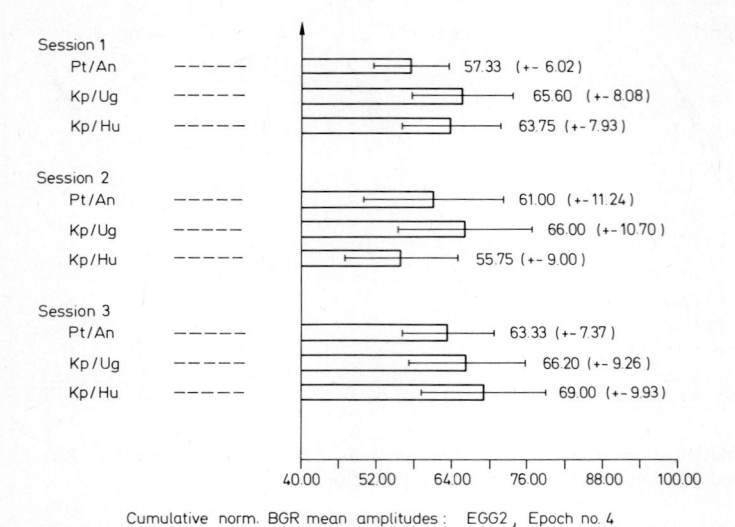

```
Session 1
   Pt/An   -----   |---[  ]---|   57.33  (+- 6.02)
   Kp/Ug   -----   |----[  ]--|   65.60  (+- 8.08)
   Kp/Hu   -----   |---[   ]--|   63.75  (+- 7.93)

Session 2
   Pt/An   -----   |----[   ]---|  61.00  (+-11.24)
   Kp/Ug   -----   |----[    ]--|  66.00  (+-10.70)
   Kp/Hu   -----   |--[   ]---|    55.75  (+- 9.00)

Session 3
   Pt/An   -----   |---[   ]--|    63.33  (+- 7.37)
   Kp/Ug   -----   |----[  ]---|   66.20  (+- 9.26)
   Kp/Hu   -----   |----[   ]--|   69.00  (+- 9.93)

          40.00  52.00  64.00  76.00  88.00  100.00
```

Cumulative norm. BGR mean amplitudes : EGG2, Epoch no. 4

Fig. 11. Changes in "time constants" of group feeding responses over three sessions. *Bars* indicate group mean with standard deviation of raw scores. *PT/AN,* patients; *KP/UG,* K1; *KP/HU,* K2. Explanation in text

Discussion

Data on gastrographic feeding responses in anorexia nervosa patients presented in this report apparently confirm Dubois' results on retarded emptying in this group. It has to be kept in mind, however, that his method of fractional emptying with an intragastric tube directly measures *emptying,* albeit of an unrealistic "meal", i.e., plain water. How serious (inhibitory?) effects of the intragastric tube on motility could influence measurements of emptying is not clear at the moment, but this has to be considered when data are compared. The method used here measures postprandial changes in *motility* by indirect surface techniques, and with a more realistic test meal. That these studies seem to replicate each other at least partially is significant. But the differences also deserve discussion.

Motility of GIT segments and the stomach in particular is not related to emptying rates in a straightforward manner. It is only one of several factors contributing to net transit rate. Except for Dubois' reports this was sometimes not considered in earlier publications (e.g., Crisp 1965, 1967). The present data therefore have to be interpreted with caution. On the other hand, emptying of water may not be a valid indicator of stomach emptying times for non-zero-calorie meals and meals of different of other consistencies. Furthermore, motility is an important variable of GI function in its own right. With the method used the study of visceral perception in anorexia is easily possible under realistic conditions.

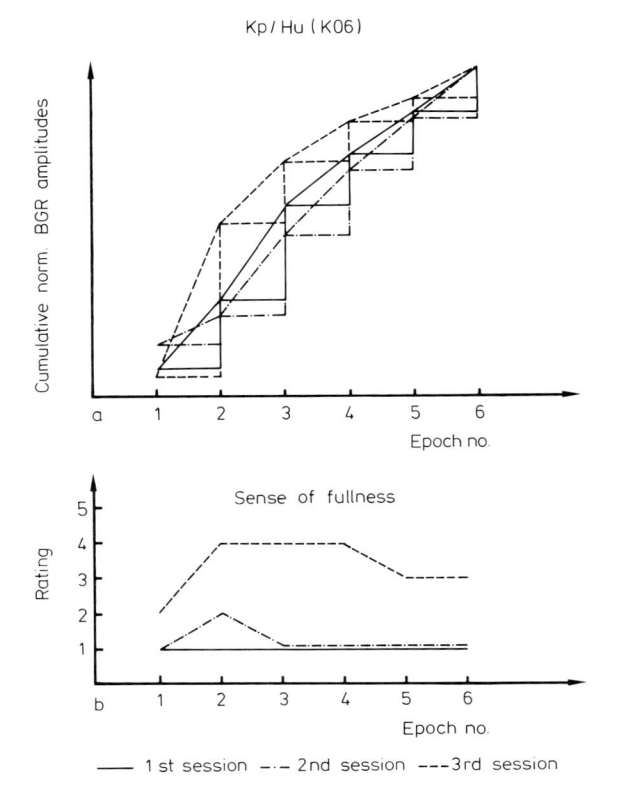

Fig. 12. Individual pattern of changes in feeding response and subjective ratings over three sessions in a dieting subject. Data from a dieting control subject (group K2, K/Hu) are displayed similar to previous figures. Weight at 1st session 93%, at 2nd session 89%, and at 3rd session 85% of normal weight. *Upper half:* Cumulative BGR amplitudes as previously explained. *Lower half:* Subjective five-point ratings of fullness. Further explanation in text

The fact that the retardation of the feeding response is diminished during later stages of treatment might suggest that it is a secondary effect of malnutrition. However, dieting volunteers do not show this retardation. On the contrary, their postprandial motility pattern is accelerated. Thus delayed motility increases after ingestion of a test meal seem to be a genuine feature of the anorectic feeding pattern. The covariation of subjective variables gives important hints for the interpretation of this feature. Figure 12 shows individual cumulative feeding response curves of a subject from group K2 together with subjective ratings of fullness from the 1st to the 3rd session. There is some similarity with the group results in that a clear acceleration of the feeding response is seen after 6 weeks of dieting. But subjective ratings show a pattern which one would expect from an anorectic patient. Individual data on subjective ratings in anorectic patients, on the other hand, do not show this plausible effect. It appears, then, that dieting control subjects with *accelerated* or back-to-normal feeding responses feel "stuffed" after a test meal of reasonable size, while anorectic patients do *not* report subjective fullness with *retarded* feeding responses! This could be just another consequence of disturbed interoception in this group, which has been postulated by several investigators in the field (cf. Garfinkel, this volume). Group analysis of subjective data on our feeding tests

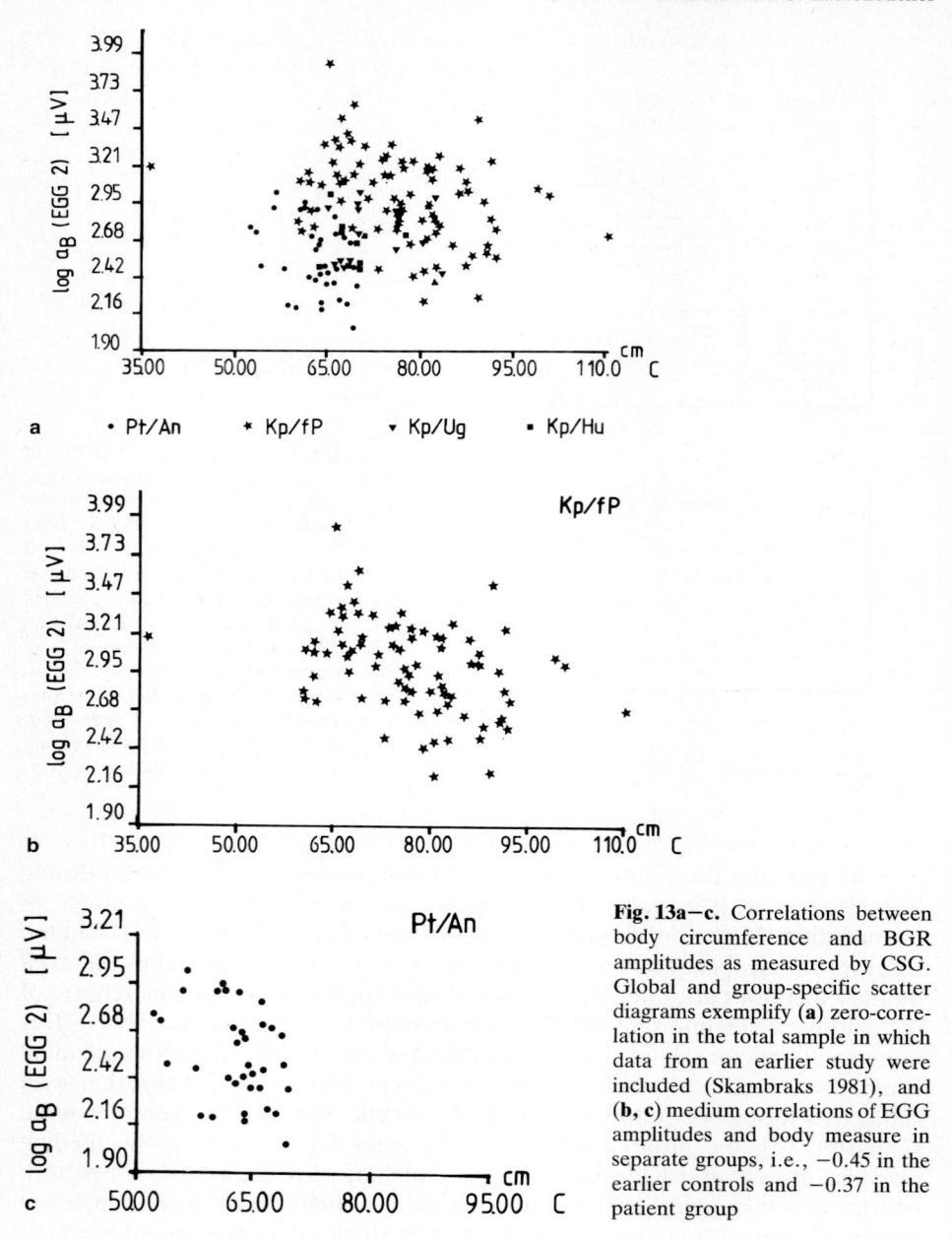

Fig. 13a–c. Correlations between body circumference and BGR amplitudes as measured by CSG. Global and group-specific scatter diagrams exemplify (**a**) zero-correlation in the total sample in which data from an earlier study were included (Skambraks 1981), and (**b, c**) medium correlations of EGG amplitudes and body measure in separate groups, i.e., −0.45 in the earlier controls and −0.37 in the patient group

should clarify this question. Because of the limited number of cases included in this preliminary report these data were not analysed here.

Whether a retarded feeding response of motility and emptying variables has major etiological significance or even relevance for risk of relapse will have to await appropriate follow-up studies. The same is true for a possible therapeutic implication of this finding: it is feasible to include retraining of normal feeding response patterns by psychophysiological techniques analogous to pharmacological attempts to speed up emptying (see *Introduction*). This might enhance effects of behavioral treatments and/or increase their persistence.

A possible artefact contaminating surface gastrographic recordings of the feeding response must be discussed at this point. In earlier studies negative correlations of EGG amplitudes and body circumference as well as thickness of abdominal fat layer were found. The correlations are only of medium height (-0.30 to -0.60), but high enough to account for some of the amplitude variance. With extremely thin abdominal walls in anorectic patients this may be an important variable, especially in repeated measurement designs with body measurements changing during weight gain. In the present study body measurements, in particular circumference and fat layer, were taken before each session. The scatter diagrams in Fig. 13 show part of the results of this correlational analysis. On the whole there seems to be no sizable dependency of gastrographic amplitudes on body measurements. But correlations within groups are significant if low. This is especially true for the anorectic patients and may influence changes in absolute response measurements by changes in body parameters during therapy. The use of relative and not absolute feeding curves such as our normalized response characteristics therefore seems advisable. In this way changes in the initial level of gastrographic amplitudes will not contaminate changes in response speed.

Confusion of primary gastrographic feeding response by vegetative correlates of emotional consequences of the meal requirements has not yet been considered at this stage. Analysis of the corresponding cardiac-respiratory group results will be reported elsewhere. A number of other problems must also be solved before one can generally recommend surface gastrography for measurement of feeding response in patients with eating disorders. While the reliability of the feeding scores may be sufficient to show significant differences even in small group designs this seems to be not enough for individual diagnostic use. Unreliability of MGGs with fasting subjects in whom frequent magnet loss into the duodenum in early phases of the feeding test can be observed presents further problems. In addition, it is difficult to relate our motility scores to emptying. A possible solution to both problems may be the use of a variant of the magnetogastrographic technique originally introduced by Frei et al. (1970). Instead of a small magnet, magnesium ferrite suspension is used as a magnetic agent. Any change in ferrite concentration during gastric emptying is measured by a susceptibility probe and used as an index of emptying. In a pilot study (cited in Hölzl 1983) we were able to show that with sufficient sensitivity of the measurement system not only emptying but also motility can be measured with this procedure, because contractions deform the ferrite bolus and give rise to transient field changes. An improved version is currently under test in our laboratory.

At present signal analysis in CSG is not well suited to detect transient events in the records like single contractions with sufficient signal-to-noise ratios. This would be needed in the study of visceral perception of patients suffering from eating disorders. The gross subjective ratings used so far can be of only limited value. The pilot study conducted by Striegel (1978) showed this very clearly. It was confirmed by the variability of subjective response patterns in the present study. Assessment of Pavlovian conditioned feeding responses, which according to the pioneering work by Hollis (1982) seem to play a major role in the regulation of eating behavior, would also demand transient detection. The so-called interdigestive migrating complex is another important transient feature of GI activity. Its inhibition seems to be an important part of the feeding response, especially of those elements controlled by humoral factors and gastrointestinal peptides (Wingate 1981; Dockray 1982). Partial solutions to these problems are conceivable to date. The psychophysiological study of eating disorders may now be integrated with established fields of study.

Acknowledgements. A number of colleagues from the MPIP have assisted the work reported here. We would like to acknowledge their help and at least mention a few. Dr. Pirke not only provided the opportunity to test the feeding responses in volunteers dieting in his endocrinological project (see references below). He also stimulated much of the present research by suggesting the use of conjoint gastrography within the MPIP Anorexia Project. Secondly we honestly admit that no single measurement of the study would have been successfully completed without Inge Riepl's psychological and technical skills. Inge's competent guidance of patients and "experimenters in charge" alike prevented impending disasters more than once if not always. She also assisted in data analysis and preparation of figures. Kurt Löffler and Gerd Müller (Department of Psychology) helped us with biosignal analysis and statistical computation. Dr. Pahl from the Department of Clinical Chemistry was responsible for arranging session dates with patients from the psychiatric wards. He also assisted in several test sessions. Lutz Erasmus and Michael Sladeczek from the Psychophysiology Lab took part as experimenters in charge and helped with the equipment.

References

Ball GG (1974) Vagotomy: effect on electrically elicited eating and self-stimulation in the lateral hypothalamus. Science 184: 484–485
Berger R (1977) Psyclosis. The circularity of experience. Freeman, San Francisco
Booth DA (ed) (1978) Hunger models. Computable theory of feeding control. Academic Press, London
Bruch H (1973) Eating disorders. Basic Books, New York
Cannon WB, Washburn AL (1912) Explanation of hunger. Am J Physiol 29: 441–454
Crisp AH (1965) Some aspects of the evolution, presentation, and follow-up of anorexia nervosa. Proc R Soc Med 58: 814–820
Crisp AH (1967) Some aspects of gastric function in anorexia nervosa. Atti 1 Congr Nazion Soc Ital Med Psicosom 3–9
Dockray GJ (1982) The physiology of cholecystokinin in brain and gut. Br Med Bull 38: 253–258
Dubois A, Gross HA, Ebert MH, Castell DO (1979) Altered gastric emptying and secretion in primary anorexia nervosa. Gastroenterology 77: 319–323
Dubois A, Gross HA, Richter JE, Ebert MH (1981) Effect of bethanechol on gastric functions in primary anorexia nervosa. Dig Dis Sci 26: 598–600
Eisman E (1966) Effects of deprivation and consummatory activity on heart rate. J Comp Physiol Psychol 62: 71–75
Frei EH, Benmair X, Yerushalmi S, Dreyfuss F (1970) Measurements of the emptying of the stomach with a magnetic tracer material. IEEE Trans Magn MAG 6: 348–349

Garfinkel PE, Moldofsky H, Garner DM, Stancer HC, Coscina DV (1978) Body awareness in anorexia nervosa: disturbances in "body image" and "satiety". Psychosom Med 40: 487–498

Hölzl R (1983) Surface gastrograms as measures of gastric motility. In: Hölzl R, Whitehead WE (eds) Psychophysiology of the gastrointestinal system: experimental and clinical aspects. Plenum, New York, pp 69–121

Hölzl R, Lautenbacher S, Brüchle H, Müller G (1983) Vegetative Reaktionen in operanten Leistungstests. In: Vaitl D, Knapp TW (Hrsg) Klinische Psychophysiologie. Beltz, Weinheim (Experimentelle Ergebnisse der Klinischen Psychologie, Bd 1)

Hollis KL (1982) Pavlovian conditioning of signal-centered action patterns and autonomic behavior: a biological analysis of function. In: Rosenblatt JS, Hinde RA, Beer C, Busnel M-C (eds) Advances in the study of behavior, vol 12. Academic Press, New York, pp 1–64

Kohatsu S (1970) The current status of electrogastrography. Klin Wochenschr 48: 1315–1319

Konturek SJ, Rösch W (1976) Gastrointestinale Motilität. In: Konturek SJ, Classen M (Hrsg) Gastrointestinale Physiologie. Witzstrock, Baden-Baden, S 3–61

Kupfermann I (1981) Hypothalamus and limbic system II: motivation. In: Kandel ER, Schwartz JH (eds) Principles of neural science. Elsevier, New York, pp 450–460

Müller GM, Hölzl R, Brüchle HA (1983) Conjoint gastrography: principles and techniques. In: Hölzl R, Whitehead, WE (eds) Psychophysiology of the gastrointestinal system: experimental and clinical aspects. Plenum, New York, pp 123–159

Pirke KM, Lemmel W, Schweiger U, Berger M, Krieg C (to be published) Influence of caloric restriction on LH and FSH secretion pattern in healthy adult women

Porges SW, Bohrer RE, Cheung MN, Drasgow F, McCabe PM, Keren G (1980) New time-series statistic for detecting rhythmic co-occurrence in the frequency domain: the weighted coherence and its application to psychophysiological research. Psychol Bull 88: 580–587

Saleh JW, Lebwohl P (1980) Meoclopramide-induced gastric emptying in patients with anorexia nervosa. Am J Gastroenterol 74: 127–132

Skambraks M (1981) Psychophysiologische und psychologische Unterschiede zwischen Patienten mit Ulcus ventriculi aut duodeni, Personen mit vegetativen Magenbeschwerden und normalen Kontrollpersonen. Thesis, University of Munich

Smout AJPM (1980) Myoelectrical activity of the stomach. Gastroelectromyography and electrogastrography. Delft University Press, Delft

Striegel R (1978) Hungerattribution. Ein psychophysiologisches Experiment zur Hungerwahrnehmung normalgewichtiger, übergewichtiger und anorektischer Personen. Thesis, University of Tübingen

Stunkard AJ, Fox S (1974) The relationship of gastric motility and hunger. Psychosom Med 33: 123–134

Wingate DL (1981) Backwards and forwards with the migrating complex. Dig Dis Sci 26: 641–666

Endocrine Function in *Magersucht* Disorders

P. J. V. Beumont[1]

Professor Ploog has posed three important questions for consideration at our symposium. These questions may be paraphrased as follows:
1. Are all the physical manifestations of anorexia nervosa a direct result of undernutrition, or are other major factors also involved?
2. If undernutrition is the prime cause, how does its presence bring about the physical symptoms that characterize the illness?
3. Is knowledge of the pathogenesis of the physical changes in anorexia nervosa merely a matter of academic interest? Or is such knowledge of practical significance, substantially increasing our understanding of the illness?

I will try to address these issues in the context of a particular endocrine dysfunction, namely that of the hypothalmic − pituitary − ovarian axis. But first I must draw the reader's attention to the behavioral aspects of the illness.

The Influence of Behavioral Disturbances on Endocrine Function

Anorexia nervosa is a complex illness and numerous factors contribute to its occurrence in the individual patient. Amongst these must be included the social pressures that exist in our culture concerning eating and not eating, abstention and gratification. Adolescent girls and young women react to these social pressures in many different ways, some of which are maladaptive and lead to illness. Anorexia nervosa necessarily involves a number of disturbances of behavior, all of which are directed towards *Magersucht*, that is to a relentless pursuit of thinness. There is, however, great variation between patients with regard to the prominence accorded to the various types of behavioral deviation.

On the basis of such variation, anorexia nervosa may be divided into a number of categories (Beumont et al. 1976b; Beumont 1977). The dieting or abstaining type closely resembles the classic picutre described by Gull (1873) and by Lasegue (1873). The vomiting and purging kind, bulimia nervosa (Russell 1979) overlaps with the syndromes of binge eating or bulimia (American Psychiatric Association 1980). More recently, attention has been focussed on a third category. Excessive exercising is a form of anorexic behavior, and

---

[1] The University of Sydney, Department of Psychiatry, Royal Prince Alfred Hospital, Camperdown, NSW 2006, Australia

The Psychobiology of Anorexia Nervosa
Edited by K. M. Pirke and D. Ploog
© Springer-Verlag Berlin Heidelberg 1984

Endocrine Function in *Magersucht* Disorders

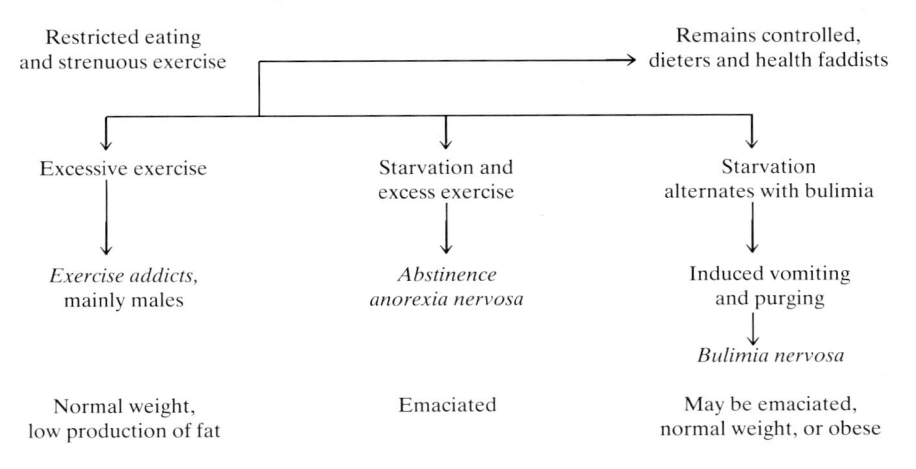

```
              Restricted eating                               Remains controlled,
           and strenuous exercise                          dieters and health faddists
                     │                                              │
         ┌───────────┼───────────┐                                  │
         ▼           ▼           ▼                                  ▼
   Excessive     Starvation and                              Starvation
    exercise    excess exercise                         alternates with bulimia
       │              │                                              │
       ▼              ▼                                              ▼
 Exercise addicts,  Abstinence                              Induced vomiting
  mainly males    anorexia nervosa                            and purging
                                                                    │
                                                                    ▼
                                                            Bulimia nervosa

  Normal weight,      Emaciated                           May be emaciated,
 low production of fat                                normal weight, or obese
```

Fig. 1. Interrelationships of the *Magersucht* disorders

obligatory running has been suggested as a further analog of the illness (Yates et al. 1983). A schematic respresentation of the various types of anorexia nervosa and how they relate to other behavioral anomalies is shown in Fig. 1. In deference to our hosts, I propose the term "*Magersucht* disorders" as rubric for the whole group.

The difficult diagnostic problem is usually not to differentiate *Magersucht* disorders from other psychiatric or physical illnesses, but rather to distinguish them from normal and perhaps even healthy patterns of living, such as persistent dieting and committed exercising. Patients differ from normal subjects not so much in the types of behavior they display, but in their *addiction* to those behaviors — the relentless manner in which they pursue them and their apparent inability to desists.

Among the three main groups of abstinence anorexics, bulimics, and obligatory exercisers, there is, of course, a great overlap. In addition to contributing to the general undernutrition, each behavior may itself directly affect endocrine function. The composition of the patient's diet is a case in point. Anorexia nervosa patients eschew energy-dense foods, avoiding fats and carbohydrates and taking a relatively high proportion of their calories in the form of protein (Beumont et al. 1981). It has long been known that carbohydrate metabolism is profoundly affected by variations in dietary composition, and Wurtman and Wurtman (this volume) have elegantly described the changes in brain neurotransmitters that follow ingestion of various normal foodstuffs. Yet published reports of studies on anorexia nervosa seldom give data concerning the immediate and long-term dietary intake of the patients. Perhaps it is assumed that their diet is normal because they have been hospitalized. Such an assumption would be unwarranted, as the compliance of anorexia nervosa patients in regard to dietary matters is notoriously bad (Naish 1979).

Strenuous exercise similarly is important in the causation of at least some of the hormonal abnormalities of anorexia nervosa. Exercise affects endocrine function, particularly with respect to carbohydrate metabolism (Soman et al. 1979). Again, exercise status is seldom mentioned in reports of endocrine dysfunction in the literature. Nor is sufficient attention given to the presence and severity of self-induced vomiting, purgation, and diuretic abuse. These behaviors effectively alter dietary intake and lead to disturbances of electrolyte balance, particularly potassium depletion, which in turn influence other physiological functions.

Perhaps Professor Ploog's first question needs to be rephrased. I suggest it should read: Can all the physical changes of anorexia nervosa be explained by undernutrition and by the behavioral disturbances that bring about that undernutrition, or is it necessary to invoke other concepts, such as stress or a predisposing hypothalmic defect? I believe that, until the effects of all the behavioral disturbances, as well as of undernutrition per se have been thoroughly documented, we should assume that they are the sole causes of the physical disorder. This certainly appears to be so in the case of the endocrine abnormalities.

Reproductive Function in Emaciation

Endocrine changes have always received prominent attention among the various physical manifestations of anorexia nervosa, and they were mentioned by Gull in his original description of the illness. Simmonds' (1918) unfortunate reference to weight loss in patients with panhypopituitarism is frequently blamed for the confusion that subsequently arose between anorexia nervosa and pituitary disease. Despite the critical appraisals of workers such as Sheldon (1939), Escamilla and Lisser (1942), and Sheehan and Summers (1948), this confusion persisted for many years. As recently as 1969, the influential Cecil and Loeb's *Textbook of Medicine* included anorexia nervosa not in the psychiatry section, but in that concerned with endocrine disease!

Over the last 15 years, understanding of the endocrine disturbance in anorexia nervosa has advanced rapidly. There is now a large body of data, the result of carefully performed research replicated by workers in many centers, demonstrating a pattern of dysfunction quite different from that associated with pituitary disease. The established facts are not in dispute. Controversy persists, however, with respect to their interpretation.

Of all aspects of endocrine dysfunction in anorexia nervosa, that relating to the menstrual disturbance has probably received most attention. A menstrual disorder is so common a finding, at least during the phase of emaciation, that the presence of normal menses leads most clinicians to seriously doubt the diagnosis (Russell 1970; Feighner et al. 1972). The disorder is basically one of anovulation.

In patients at low body weights, circulating values of LH (luteinizing hormone) are depressed. LH levels are undoubtedly related to the severity of the emaciation, and as weight is regained during refeeding, circulating LH is restored to normal tonic values. The critical weight for the restoration of normal levels is about 80% of standard body weight, i.e., the average weight of individuals of the same age, sex, and height (Beumont et al. 1973, 1976a).

The cause of the low LH values is not to be found in the pituitary. If the gland is stimulated by a course of several days' administration of LHRH (LH releasing hormone) normal values of LH are restored without any necessary weight gain (Nilius and Wide 1975; Yoshimoto et al. 1975). This finding suggests that the primary fault may be inadequate stimulation of the pituitary gonadotrophs by LHRH, and places the site of disturbance back to the hypothalamic level.

The dynamics of the dysfunction are illustrated by the use of various provocative tests of the hypothalmic−pituitary−ovarian axis. The inhibiting effect of ethinylestrodiol on LH can not be discerned at very low body weights (Wakeling et al. 1977). Clomiphene, an antiestrogen which blocks the inhibiting feedback of endogenous estrogen, causes by this means a rise in LH values. In emaciated anorexia nervosa patients the effect is minimal or nonexistent (Marshall and Fraser 1971; Beumont et al. 1973). Further, the stimulatory effect of a bolus or short-term infusion of exogenous LHRH is markedly depressed in these subjects (Beumont et al. 1976a).

The simplest explanation of these findings is that the pituitary gonadotrophs undergo disuse atrophy as a result of inadequate LHRH stimulation during emaciation. Confirmation is given by the report of a selective degeneration of pituitary gonadotrophs in a postmortem study by Halmi (1976).

As the patient is refed and regains weight, normal tonic levels of LH are restored. Ethinylestradiol once more exerts its inhibitory effect (Wakeling et al. 1977), clomiphene produces a short-term rise in LH (Beumont et al. 1973), and LHRH may even evoke an exaggerated LH response reminiscent of that seen during puberty (Beumont et al. 1976a).

Plasma FSH (follicle-stimulating hormone) levels tend to follow a similar pattern to those of LH. They are often depressed in patients at very low weight, but values are restored to normal by increases in weight smaller than those associated with the restoration of LH levels (Beumont et al. 1976a). Similarly, the FSH response to LHRH becomes exaggerated at an earlier phase than the LH response (Beumont et al. 1976a; Warren 1977), a phenomenon which accounts for the finding of an "inverted" gonadotrophin response − high FSH, low LH − that has been reported in some studies of anorexia nervosa patients (Nilius et al. 1975; Sherman et al. 1975). Findings on FSH, however, are less consistent than those on LH. Some studies have reported normal tonic values of FSH and an unremarkable response to LHRH (Beumont and Abraham 1981). This is not surprising when we remember that the control of FSH remains poorly understood, and that there is some doubt as to whether LHRH is the major stimulant of its release (Pasteels and Franchimont 1976).

Delayed Resumption of the Menses

The association between undernutrition and decreased circulating levels of gonadotrophins is not sufficient to account for all aspects of the menstrual disorder of anorexia nervosa. Although tonic levels of LH are restored by weight gain, the cyclical changes of hormones that herald ovulation are usually long delayed.

Frisch (1977), on the basis of her work concerning the role of body fat in determining the timing of menarche, has postulated that a critical content of fat in body composition has not yet been obtained in those anorectic patients whose amenorrhea persists after weight restoration. Estrogens are metabolized in fat tissue. If fat is inadequate, estrogens which are less active biologically may predominate. These would be ineffective in their central feedback on the hypothalamus. Support for this theory derives from the findings of Jeuniewic et al. (1978) of inadequate weight/height ratios in anorexia nervosa patients after treatment. There is also indirect supportive evidence from a number of studies of apparently healthy women who show anovulation in the presence of a disproportionately low fat content in body composition, e.g., athletes and ballet dancers.

Frisch's ideas have been criticized and alternative explanations have been advanced for the associations she observed. In respect to the frequent occurrence of menstrual disorders in athletes and in ballet dancers, it has been suggested that *stress* may play an important role (Smith 1980), or that a *selection factor* may operate which induces women with potentially poor menstrual function to opt for strenuous exercise (Malina et al. 1978). The physique of late-maturing girls is characterized by longer legs and narrower hips, an advantage for athletic pursuits.

In ballet dancers, a relationship has been demonstrated between intense physical exercise and amenorrhea in the presence of weight/height ratios that appear adequate (Abraham et al. 1982). If training is interrupted in these subjects, menses return without alteration of body weight or the lean/fat ratio. These data suggest that strenuous exercise per se, rather than a critical decrease in fat tissue, may be the factor responsible, but the mechanism involved remains obscure.

Inadequate weight/height ratios and excessive exercising commonly persist in anorexia nervosa patients after weight restoration, and may contribute to the usual delay in the return of normal menstrual function. There appears, however, to be yet another factor involved.

During the normal menstrual cycle, estrogen exerts two opposing actions on circulating gonadotrophins. Up to a critical threshold it depresses LH levels. Once this threshold is exceeded, as it is just prior to ovulation, estrogen increases pituitary capacity and actually induces a rise in LH values (Yen 1977). Similar changes can be noted following the administration of exogenous ethinylestradiol and of clomiphene. In many weight-restored patients the negative feedback effects of these drugs can be evoked, but not their positive effects (Wakeling et al. 1977; Beumont et al. 1973). These findings suggest a persistent insensitivity of the relevant hypothalamic mechanism.

Further information about this insensitivity has emerged from studies using a 4-hour infusion of LHRH. In emaciated subjects no response is observed, which is compatible with the view that the pituitary, at this stage, contains virtually no LH. Following weight restoration an LH peak occurs early in the course of infusion, indicating the presence of a readily accessible LH pool, but maintained stimulation does not evoke a further response. Only after full rehabilitation is a normal biphasic response obtained, showing the restoration not only of the readily accessible pool of LH, but also of a second or deep pool within the pituitary (Beumont and Abraham 1981). Yen (1977) suggests that the first LH pool is dependent entirely on priming by LHRH, and that the second pool is determined by the positive feedback effects of estrogen. Its deficiency in inadequately rehabilitated anorectic patients indicates defective estrogen feedback at the pituitary as well as at the hypothalamic level.

Early Onset of Amenorrhea

Related to the problem of the persistence of amenorrhea following weight restoration is the question of the early onset of menstrual disorder in anorexia nervosa patients. Menstrual disturbance often antedates the other obvious manifestations of the illness, sometimes by many months (Decourt 1953). A number of theories have been put forward to explain this observation. It has been suggested, for instance, that girls with unstable menstrual function are more likely to become anorectic. This appears unlikely in view of reports of an early age of menarche in subjects who subsequently develop anorexia nervosa (Crisp 1967). After all, early menarche usually portends a long and stable menstrual history.

Stress has also been invoked as an explanation, and an analogy drawn with the so-called psychogenic amenorrheas observed both in war time and in peace. The literature in this area is rather confused. Studies undertaken in a war-time setting do not provide unequivocal evidence of psychogenic causation, since the stress of internment or siege seldom occurred in the absence of some nutritional deprivation (Russell 1972). Studies in peace time also have their deficiencies. First, there are reports of missed periods after acute psychological trauma at around the time of ovulation (Rakoff 1968). These sound genuine enough. Second, there are instances of psychotic patients having amenorrhea. In some instance at least, neuroleptic-induced hyperprolactinemia may be the cause of their menstrual difficulties (Beumont et al. 1974). Third, there are references to neurotic women and girls with long-standing emotional problems and menstrual disorders. Are they undiagnosed cases of anorexia nervosa? In any case, none of these examples appears to further our understanding of the pathogenesis of the disturbance.

The case for stress as a cause of the early amenorrhea is at best unproven. There is in fact little evidence that gonadotrophins are stress-dependent hormones, although adrenocorticotrophin, growth hormone, and prolactin undoubtedly are (Guevara et al. 1970).

A third possible explanation is that variations in diet that antedate weight loss bring about the early occurrence of menstrual disorder. In rats, Meites and Reed (1949) found that a low-calorie diet in which calorie intake was not reduced sufficiently to cause weight loss brought about anestrus within 3 weeks. Dieting causes a temporary menstrual disturbance in obese women, with lowered gonadotrophin levels (Beardwood 1974). A similar phenomenon may be associated with the onset of dieting in anorexia nervosa. It is difficult to date the beginning of the behavioral disturbance in these patients, but the obvious manifestations of the illness — emaciation, depression, and preoccupation with dieting — usually occur after months or even years of progressive dietary restriction (Beumont et al. 1979).

Dietary restriction leads to a reduction in serum glucose and a lowered blood insulin concentration (Malaisse et al. 1967). Ibrahim and Howland (1972) suggested that the hypothalamic area responsible for the synthesis and release of LHRH is exquisitely sensitive to variations in insulin concentration, and in the presence of too low a level LHRH release is suppressed.

Other Hormones

Over the last two decades, the pathogenesis of most hormonal disturbances in anorexia nervosa patients has been unravelled and the dysfunction explained in purely physiological terms. Two examples will serve to illustrate this point. First, the elevated growth hormone levels which occur in about 60% of patients (Marks et al. 1965) are associated with inadequate caloric intake in the current diet (Frankel and Jenkins 1975; Brown et al. 1977). The abnormality is promptly corrected by restoration of a normal nutritional intake (Casper et al. 1977). Second, low normal T_4 levels and markedly decreased T_3 levels are found in emaciated patients (Moshang and Utiger 1977) as in many other situations, such as chronic illness, protein-calorie malnutrition, and the prolonged starvation of obese subjects. The phenomenon has been shown to derive from a change in the peripheral metabolism of thyroid hormone (Chopra et al. 1975). T_4 may be degraded peripherally either to 3,5,3'-triiodothyronine (T_3) or to 3,3',5-triiodothyronine, reverse T_3, which is biologically far less active. In catabolic conditions, the reverse T_3 pathway is favored, presumably as a means of conserving energy (Vagenukis 1977).

In some instances, the endocrine dysfunction persists after weight has been restored and the patient has apparently recovered. An example is the disturbance affecting the central release of vasopressin (Ebert et al., this volume). Does this necessarily indicate a disorder independent of the undernutrition and the behaviors that bring it about? Not so. The rate of recovery following nutritional deficiency varies greatly between systems. Further, the resumption of a normal body weight does not necessarily indicate that the patient has completely abandoned all the behaviors that had made her ill. In many instances, these malignant practices continue for years after apparent recovery.

Conclusion

It is clear that a great deal is known about the pathogenesis of endocrine dysfunction in anorexia nervosa. There remain, of course, some major gaps in our knowledge, but even here hypotheses can be propounded which sound reasonable and are testable by experimental means. It seems that a satisfactory answer can be given to Professor Ploog's second question: we do have a fairly good idea how undernutrition and the behaviors with which it is associated bring about the physical symptoms characteristic of the illness, at least those arising from the endocrine system.

But does this knowledge substantially increase our understanding of the illness? I think not. The physical symptoms of anorexia nervosa are epiphenomena unrelated to the etiology of the illness. It is useful to know about them — but mainly in order to avoid unnecessary investigation of the patient (Lancet 1979). Of course, the presence of physical disorder does eventually affect the patients behavior, and to some extent, a vicious circle of illness may be set up. But the many elaborate schemes of interacting variables that have been proposed, such as that suggested by Russell (1977), do not really *explain* the illness. Anorexia nervosa remains basically a neurotic disturbance, leading to deviations in behavior that bring about physical dysfunction. Its treatment, albeit unsatisfactory, is psychological.

References

Abraham SF, Beumont PJV, Fraser IS, Llewellyn-Jones D (1982) Body weight, exercise and menstrual status among ballet dancers in training. Br J Obstet Gynaecol 89: 507–510
American Psychiatric Association (1980) Diagnostic and statistical manual of mental disorders, 3rd ed. American Psychiatric Association, Washington
Beardwood CJC (1974) Relationship between body weight and gonadotrophin excretion in anorexia nervosa and obesity. S Afr Med J 48: 53–58
Beumont PJV (1977) Further categorization of patients with anorexia nervosa. Aust N Z J Psychiatry 11: 223–226
Beumont PJV, Abraham SF (1981) Continuous infusion of luteinizing hormone releasing hormone (LHRH) in patients with anorexia nervosa. Psychol Med 11: 477–484
Beumont PJV, Carr PJ, Gelder MG (1973) Plasma levels of luteinizing hormone and of immunoreactive oestrogens (oestradiol) in anorexia nervosa: response to clomiphene citrate. Psychol Med 3: 495–501
Beumont PJV, Abraham SF, Argall WJ (1979) Variables of onset in anorexia nervosa. In: Obiols J, Ballus C, Gonzales Monclus E, Pujol J (eds) Biological psychiatry today. Elsevier North-Holland Biomedical Press, Amsterdam Oxford New York, pp 597–601
Beumont PJV, Corker CS, Friesen HG, Gelder MG, Harris GW, Kolakowska T, MacKinnon PCB, Mandelbrote BM, Marshall J, Murray MAF, Wiles DH (1974) The effect of phenothiazines on endocrine function. Br J Psychiatry 124: 413–430
Beumont PJV, George GCW, Pimstone BL, Binik AI (1976a) The pituitary response to hypothalamic releasing hormones in patients with anorexia nervosa. J Clin Endocrinol Metabol 43: 487–496
Beumont PJV, George GC, Smart DE (1976b) 'Dieters' and 'vomiters and purgers' in anorexia nervosa. Psychol Med 6: 617–622
Beumont PJV, Chambers TL, Rouse L, Abraham SF (1981) Diet composition and nutritional knowledge of patients with anorexia nervosa. J Hum Nutr 35: 265–273
Brown GM, Garfinkel PE, Jeuniewic N, Moldofsky H, Stancer HC (1977) Endocrine profiles in anorexia nervosa. In: Vigersky RA (ed) Anorexia nervosa. Raven, New York, pp 123–136
Casper RC, Davis JM, Pandey GN (1977) The effect of nutritional status and weight changes on hypothalamic function tests in anorexia nervosa. In: Vigersky RA (ed) Anorexia nervosa. Raven, New York, pp 137–147

Chopra IJ, Chopra U, Smith SR, Reza M, Solomon DH (1975) Reciprocal changes in serum concentration of 3,3′,5-triiodothyronine (reverse T_3) and 3,5′, 3′-triiodothyronine (T_3) in systemic illness. J Clin Endocrinol Metabol 41: 1043–1049

Crisp AH (1967) Anorexia nervosa. Br J Hosp Med 1: 713–718

Decourt J (1953) Die anorexia nervosa: Psycho-endokrine Kachexie der Reifungszeit. Dtsch Med Wochenschr 78: 1619–1622

Escamilla RF, Lisser H (1942) Simmonds' disease. J Clin Endocrinol Metabol 2: 65–96

Feighner JP, Robins E, Guze SB, Woodruff RA, Jr Winokur G, Munoz R (1972) Diagnostic criteria for use in psychiatric research. Arch Gen Psychiatry 26: 57–63

Frankel RJ, Jenkens JS (1975) Hypothalamic-pituitary function in anorexia nervosa. Acta Endocrinol (Kbh) 78: 209–221

Frisch RE (1977) Food intake, fatness and reproductive ability. In: Vigersky R (ed) Anorexia nervosa. Raven, New York, pp 149–162

Guevara A, Luria MH, Wieland RE (1970) Serum gonadotrophin levels during medical stress. Metabolism 19: 79–83

Gull WW (1873) Anorexia hysterica (apepsia hysterica). Br Med J 2: 257–528

Halmi KA (1976) Selective pituitary deficiency in anorexia nervosa. In: Sachar E (ed) Hormones, behaviour and psychopathology. Raven, New York, pp 285–290

Jeuniewic N, Brown GM, Garfinkel PE, Moldofsky H (1978) Hypothalamic function as related to body weight and body fat in anorexia nervosa. Psychosom Med 40: 187–198

Ibrahim EA, Howland BE (1972) Effect of starvation on pituitary and serum FSH and LH following ovarectomy in the rat. Can J Physiol Pharmacol 50: 768–773

Kaufman MR, Heiman M (1964) Evolution of psychosomatic concepts. Anorexia nervosa: a paradigm. International Universities Press, New York

Lancet (1979) Anorexia nervosa: to investigate or to treat. Lancet 2: 563–564

Lasegue C (1873) De l'anorexie hysterique. Reprinted (1964) in: Kaufman MR, Herman M (eds) Evolution of a psychosomatic concept: anorexia nervosa. International Universities Press, New York, pp 141–155

Malaisse WJ, Malaisse-Lagae F, Wright PH (1967) Effect of fasting on insulin secretion in the rat. Am J Physiol 213: 843–848

Malina RM, Spirduso WW, Tate C, Baylor AM (1978) Age at menarche and selected menstrual characteristics in athletes. Med Sci Sports 10: 218–222

Marks V, Howorth N, Greenwood DFC (1965) Plasma growth-hormone levels in chronic starvation in man. Nature 208: 686–687

Marshall JC, Fraser TG (1971) Amenorrhoea in anorexia nervosa: assessment and treatment with clomiphene citrate. Br Med J 4: 590–592

Meites J, Reed JO (1949) Effects of restricted food intake in intact and ovarectomized rats on pituitary lactogen and gonadotrophin. Proc Soc Exp Biol Med 70: 513–516

Moshang T, Utiger RD (1977) Low triiodothyronine euthyroidism in anorexia nervosa. In: Vigersky RA (ed) Anorexia nervosa. Raven, New York, pp 263–270

Naish JM (1979) Problems of deception in medical practice. Lancet 2: 139–142

Nilius SJ, Fries H, Wide L (1975) Successful induction of follicular maturation and ovulation by prolonged treatment with LH-releasing hormone in women with anorexia nervosa. Am J Obstet Gynecol 122: 921–928

Pasteels JL, Franchimont P (1976) The production of FSH by cell cultures of fetal pituitary. In: Hubinat P (ed) Progress in reproductive endocrinology. Karger, Basel

Rakoff AE (1968) Endocrine mechanisms in psychogenic amenorrhoea. In: Michael RP (ed) Endocrinology and human behaviour. Oxford University Press, London, pp 139–160

Russell GFM (1970) Anorexia nervosa: its identity as an illness and its treatment. In: Price DH (ed) Modern trends in psychological medicine. Butterworths, London, pp 131–164

Russell GFM (1972) Psychological and nutritional factors in disturbances of menstrual function and ovulation. Postgrad Med J 48: 10–13

Russell GFM (1977) The present status of anorexia nervosa (editorial). Psychol Med 7: 363–367

Russell GFM (1979) Bulimia nervosa, an ominous variant of anorexia nervosa. Psychol Med 9: 429–448

Sheehan HL, Summers VK (1948) The syndrome of hypopituitarism. Q J Med 18: 319–398

Sheldon JH (1939) Anorexia nervosa. Proc R Soc Med 32: 738–740
Sherman BM, Halmi KA, Zamudio R (1975) LH and FSH response to gonadotrophin-releasing hormone in anorexia nervosa: effect of nutritional rehabilitation. J Clin Endocrinol Metabol 41: 131–142
Simmonds M (1918) Atrophie des Hypophysisvorderlappen und hypophysare Kachexie. D Med Wochenschr 44: 852–854
Smart DE, Beumont PJV, George GCW (1976) Some personality characteristics of patients with anorexia nervosa. Br J Psychiatry 128: 57–60
Smith NJ (1980) Excessive weight loss and food aversion in athletes simulating anorexia nervosa. Pediatrics 66: 139–142
Soman VR, Koivisto VA, Deibert D, Felig P, Defronzo RA (1979) Increased insulin sensitivity and insulin binding to monocytes after physical training. N Engl J Med 301: 1200–1204
Vagenukis AG (1977) Thyroid hormone metabolism in prolonged experimental starvation in man. In: Vigersky RA (ed) Anorexia nervosa. Raven, New York, pp 243–254
Wakeling A, de Souza VA, Beardwood CJ (1977) Assessment of the negative and positive feedback effects of administered oestrogen on gonadotrophin release in patients with anorexia nervosa. Psychol Med 7: 397–405
Warren MP (1977) Weight loss and responsiveness to LH-RH. In: Vigersky RA (ed) Anorexia nervosa. Raven, New York, pp 189–198
Yates A, Leehey K, Shisslak CM (1983) Running – an analogue of anorexia? N Engl J Med 308: 251–255
Yen SSC (1977) Neuroendocrine aspects of the regulation of cyclic gonadotrophin release in women. In: Martini L, Besser GM (eds) Clinical endocrinology. Academic, New York, pp 175–196
Yoshimoto Y, Moridera K, Imura H (1975) Restoration of normal pituitary gonadotrophin reserve by administration of luteinising hormone releasing in patients with hypogonadotropic hypogonadism. N Engl J Med 292: 242–245

Hypothalamic Pituitary Function in Starving Healthy Subjects

M. M. Fichter[1] and K. M. Pirke[2]

A large amount of literature has emerged describing endocrine dysfunctions in anorexia nervosa. Most hypothalamic-pituitary functions appear to be abnormal in the acute stage of the illness (cf. synopsis by Vigersky 1977; Beumont and Russell 1982; Garfinkel and Garner 1982). There are many indications that hypothalamic dysfunctions in anorexia nervosa are a consequence of reduced calorie intake or weight loss. However, it cannot be ruled out that these dysfunctions are secondary to emotional disturbances or that a primary biological dysfunction is responsible for the changes in hormonal secretion. The present study addresses itself to the issue of specificity of the endocrine disturbances which have been described for anorexia nervosa. If the same endocrine disturbances as in anorexia nervosa can be reproduced in healthy human beings during starvation, the hormonal changes could be considered specific for reduced calorie intake or weight loss rather than for anorexia nervosa. Therefore we studied healthy volunteers before, during, and after starvation. The results may well be relevant not only for anorexia nervosa but also for other disorders associated with changes in nutrition, such as depression and alcoholism.

Method and Experimental Design

Five healthy female subjects aged 21–25 years (23.2 ± 2.2) were thoroughly screened by psychological and laboratory tests. The study was approved by the Ethics Committee of the University of Munich; informed consent was obtained from the subjects, who had volunteered for the study. As shown in Fig. 1, subjects were studied longitudinally in four experimental phases. During initial baseline (A) the subjects kept their weight stable at 58 ± 4.9 kg (mean ± SD). During the fasting phase (B) the subjects remained permanently in our research unit to ensure close medical observation. During an average of 21 days the subjects lost about 8 kg [from 102.9% ± 2.7% of IBW (ideal body weight) to 87.8% ± 1.7% of IBW]. In the weight gain phase (C) original body weight was restored (101.8% ± 2.3% of IBW). In the final baseline phase (D) subjects kept their body weight stable. Before fasting (I), after fasting (II), and after weight gain (III) blood samples were obtained over 24 h at 30-min intervals. The 24-h secretory patterns for cortisol and the pituitary hormones (a) luteinizing hormone (LH), (b) growth hormone (HGH), (c) thyrotropin (TSH), and (d) prolactin (PRL) were determined by radioimmunoassay. When blood samples for the 24-h hormonal profiles were obtained, the following tests of endocrine functioning were performed: (a) Dexamethasone suppression test (DST) with a single oral dose of 1.5 mg dexamethasone at 11 p.m.; (b) stimulation of TSH with 200 µg protirelin (TRH); and (c) stimulation of HGH with 0.15 mg

---

1 Psychiatrische Klinik der Universität München, Nußbaumstraße 7, D-8000 München 2
2 Max-Planck-Institut für Psychiatrie, Kraepelinstraße 10, D-8000 München 40

The Psychobiology of Anorexia Nervosa
Edited by K. M. Pirke and D. Ploog
© Springer-Verlag Berlin Heidelberg 1984

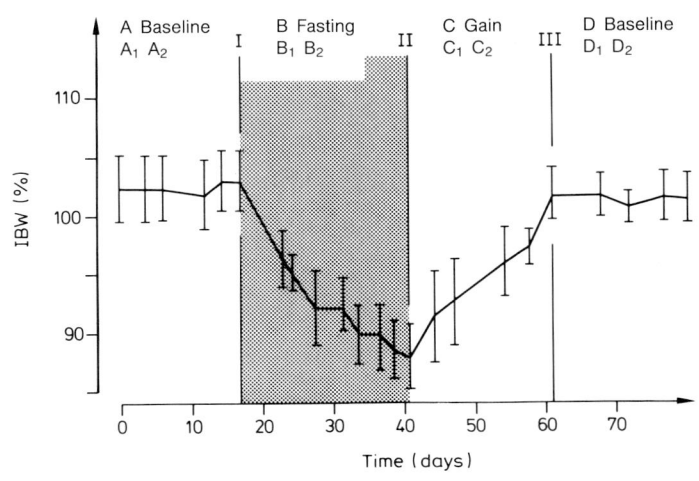

Fig. 1. Experimental design and course of body weight over time in five healthy volunteers during four experimental phases (mean and standard deviation). *IBW*, ideal body weight

clonidine in 10 ml NaCl given by slow i.v. injection. The occurrence of hypotension during starvation caused the clonidine test to be omitted in two of the five subjects. The DST and TRH tests were performed two or three times in each experimental phase (A–D). The tests for endocrine functioning were never performed on consecutive days, so that interference of effects was avoided.

Nonparametric statistical analysis was performed whenever feasible. The Wilcoxon test for paired ranks was used whenever $n = 5$ and if not indicated otherwise. Probability (P) values reported are for one-sided testing. In the 24-h secretory patterns the area under the curve was calculated using the Simpson formula, and statistical analysis was based on these data.

Results

Hypothalamic-Pituitary-Adrenal Function

Dexamethasone Suppression Test

All 11 DSTs performed during baseline phase I (A) showed normal suppression of cortisol (below 5 µg/100 ml). At 4 p.m. the day after dexamethasone application the cortisol levels (µg/100 ml) were 1.1 ± 0.1 in phase A_1 and 1.3 ± 0.1 in phase A_2 (mean and standard error); a rise of plasma cortisol was observed during starvation, to 6.1 ± 3.2 µg/100 ml in phase B_1 and 4.7 ± 1.4 µg/100 ml in phase B_2 (d.f. = 4, $P < 0.05$ for all comparisons A vs B) (Table 1). The *9 p.m.* cortisol levels after dexamethasone application results were basically the same as those at 4 p.m. During the fasting phase (B) half the DSTs (7 of 14) showed insufficient cortisol suppression from 9 a.m. to 9 p.m. following dexamethasone administration. During the weight gain phase (C) all cortisol levels after stimulation were again below 5 µg/100 ml. The postsup-

Table 1. Mean plasma levels and SE for (1) cortisol following suppression with dexamethasone; (2) TSH following stimulation with protirelin (TRH); and (3) increment ($\bar{x}$) in plasma growth hormone (HGH) following stimulation with clonidine during the four phases of the study

		Experimental phases									A_1 vs B_1	A_2 vs B_2	B_1 vs C_1	B_2 vs C_2
		Baseline I		Fasting		Gain		Baseline II						
		A_1	A_2	B_1	B_2	C_1	C_2	D_1	D_2					
1) Dexamethasone → cortisol														
4 p.m. post suppression	$\bar{x}$ SE	1.1 0.1	1.3 0.1	6.1 3.2	4.7 1.4	1.8 0.0	1.2 0.2	0.9 0.8	2.2 2.3		<0.05	<0.05	n.s.	—
9 p.m. post suppression	$\bar{x}$ SE	1.1 0.8	1.1 0.1	3.4 0.6	4.4 1.0	2.2 0.5	1.7 1.4	1.1 0.7	1.2 0.2		<0.05	<0.05	n.s.	n.s.
2) TRH → TSH														
Basal	$\bar{x}$ SE	1.6 0.8	2.1 1.5	1.0 0.5	1.1 0.3	1.5 0.8	2.5 1.2	1.2 0.5	1.0 0.3		<0.025	<0.05	<0.025	<0.05
20 min post stimulation	$\bar{x}$ SE	12.8 3.6	14.4 6.2	8.5 3.5	8.7 3.3	14.0 3.9	14.7 4.5	12.7 5.2	11.1 3.3		<0.01	<0.05	<0.025	<0.025
3) HGH														
Stimulation with clonidine ($n = 3$)	$\bar{x}$ SE		13.5 6.9		16.0 3.9		4.2 1.8					n.s.		<0.05

Wilcoxon test; d.f. = 4; ** t-test for dependent samples

pression cortisol levels at 4 p.m. had normalized to 1.8 ± 0.0 μg/100 ml in phase C_1 and 1.2 ± 0.2 μg/100 ml in phase C_2. During final baseline phase (D) 9 of 10 DSTs (29 of 30 cortisol probes between 9 a.m. and 9 p.m.) were normal (below 5 μg/100 ml).

Plasma Cortisol Secretion

The 24-h plasma cortisol secretion showed the following changes during the starvation period (mean and standard deviation): (a) The average 24-h plasma cortisol level showed a significant rise from 7.33 ± 4.2 to 12.41 ± 2.9 μg/100 ml (d.f. = 4, $P < 0.05$). After weight gain the plasma cortisol levels normalized again to 6.75 ± 4.2 μg/100 ml (d.f. = 4, $P < 0.05$); (b) the number of secretory episodes in 24 h increased during the starvation period (d.f. = 4, not statistically significant) and the total amount of time spent in secretory activity was longer after starvation (d.f. = 4, $P < 0.05$); and (c) the cortisol half-life in plasma increased during the starvation phase (d.f. = 4, $P < 0.05$).

A lack of normal responsivness to dexamethasone suppression has been reported in anorexia nervosa, suggesting a defect in the negative feedback mechanism. Basically the same findings concerning the HPA axis as we have found in starving healthy subjects have been reported in the acute phase of anorexia nervosa. Thus an increase in the cortisol production rate, a prolonged plasma half-life of cortisol, an elevation of the 24-h cortisol plasma level, and disturbances in the 24-h rhythm have been described in anorexia nervosa (Boyar et al. 1977; Doerr et al. 1980; Walsh et al. 1981). Results of our study confirm the hypothesis that starvation is one factor which results in major disturbances of the pituitary adrenal regulation.

Plasma Luteinizing Hormone

Before the fasting period all five subjects had a normal adult LH secretory pattern as defined by Pirke et al. (1979). At the end of the fasting period two subjects (3 and 5) continued to have adult LH secretory patterns, while three subjects (1, 2, and 4) showed regression to infantile LH secretory patterns. Figures 2 and 3 show 24-h LH patterns typical for each group of subjects before fasting, after fasting, and after weight gain. Data recorded in this fasting experiment with a limited number of subjects and weight loss within a limited range indicate that age, absolute weight in kilograms before fasting, and amount of weight loss (in kilograms or as percentage) during fasting do not explain the differential reaction in LH secretion of subjects during starvation. It is most likely relevant, in what phase of the menstrual cycle a subject began to fast. Those subjects who showed regression in the LH pattern began fasting later in their menstrual cycle (12th, 23rd, and 23rd days) than those who showed no change in LH secretion (4th and 5th days). It must be emphasized that the infantile secretory pattern in three subjects was observed following relatively little weight loss (9 kg at the most), within a relatively short time period (14−23 days), and at a weight level which was close to normal. In our five subjects the average weight after starvation (as percentage of IBW) was even slightly higher

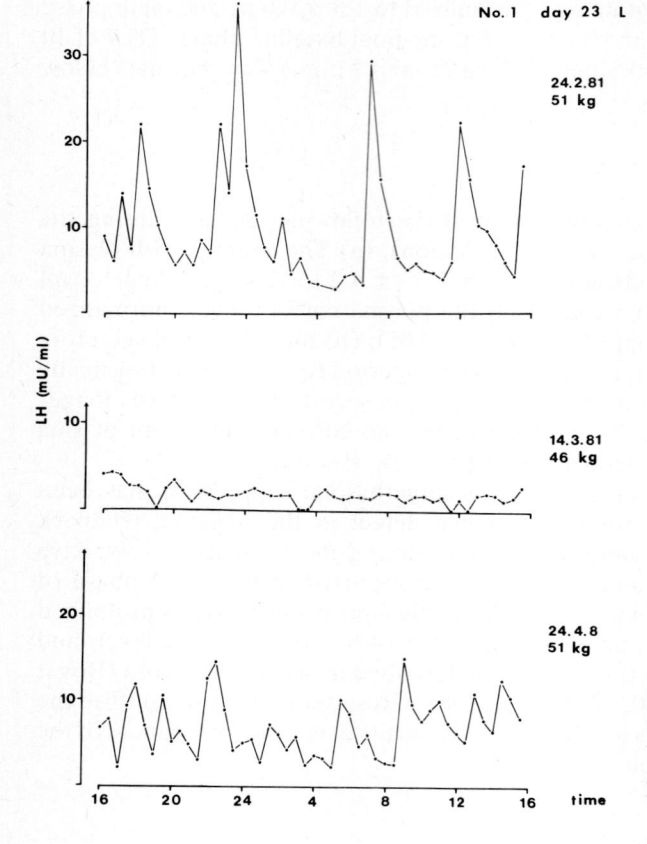

Fig. 2. 24-h LH pattern before fasting, after fasting, and after weight gain in subject 1 (regression to infantile LH pattern)

than in a series of anorexia nervosa patients, with adult LH patterns after restoration of weight (Fichter et al. 1982). In our starvation experiment one subject (5) had a menstrual period at the beginning of the fasting phase (days 5−8); other than that, all subjects were amenorrhoic during fasting, weight gain, and baseline II.

Since amenorrhea is one of the symptoms of anorexia nervosa, a disproportionate amount of research has accumulated on this topic. Low levels of gonadotropin, low levels of gonadal hormones in plasma, a diminished response to clomiphen, a disturbed positive feedback after administration, of estrogen, a delayed LH response to LH-RH administration, and disturbances in the 24-h LH secretory pattern have been described in anorexia nervosa (Boyar et al. 1974; Pirke et al. 1979). We have shown in fasting healthy subjects during starvation that even a relatively slight weight loss (compared with that in anorexia nervosa patients) can result in a regression of the 24-h LH secretory pattern. Our results with healthy starving subjects do not support the notion of absolute weight thresholds to which infantile, pubertal, and adult LH patterns (and amenorrhea) are related. Our results rather tend to confirm a hypothesis of sliding set points influenced by age, body weight history, and nutritional intake.

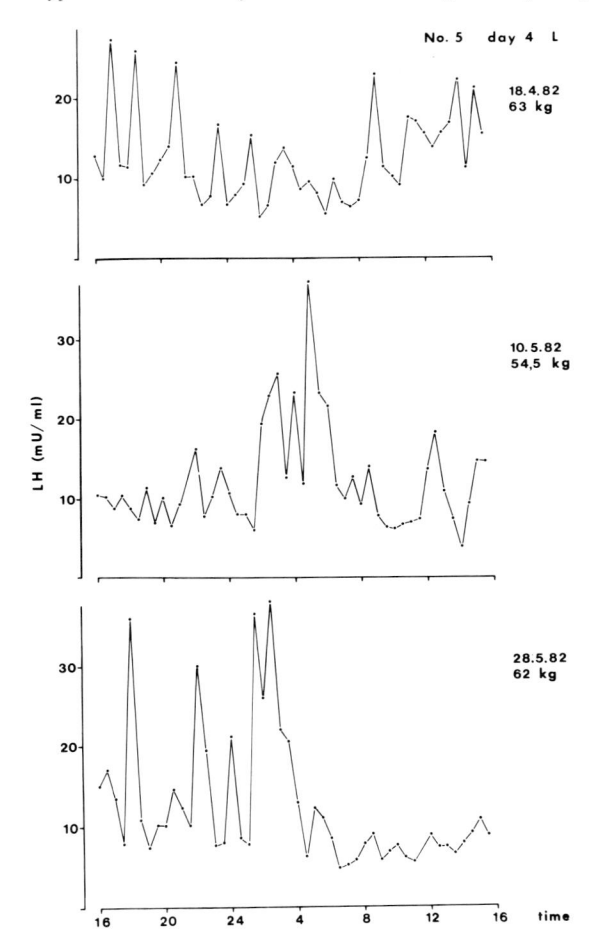

Fig. 3. 24-h LH pattern before fasting, after fasting, and after weight gain in subject 5

Future research will have to clarify the role of other factors (besides age and reduced food intake and weight loss) that influence LH secretion.

Pituitary Thyroid Function

No major difference in the *24-h TSH secretory pattern* was detected in starving healthy subjects before fasting, after fasting, or after weight gain. The mean TSH plasma level (microunits/ml) in 24 h (mean and standard deviation) was 3.16 ± 0.8 before fasting, 2.74 ± 0.6 after fasting, and 2.85 ± 0.5 after weight gain (d.f. = 3, n.s.). Figure 4 shows the 24-h TSH secretion before starvation, after starvation, and after weight gain.

In the *TRH test* a significant effect of starvation on the TSH basal plasma levels (9 a.m.) and the TSH plasma levels 30 min after stimulation with TRH was observed (d.f. = 3, *t*-test for paired observations, $P < 0.05$).

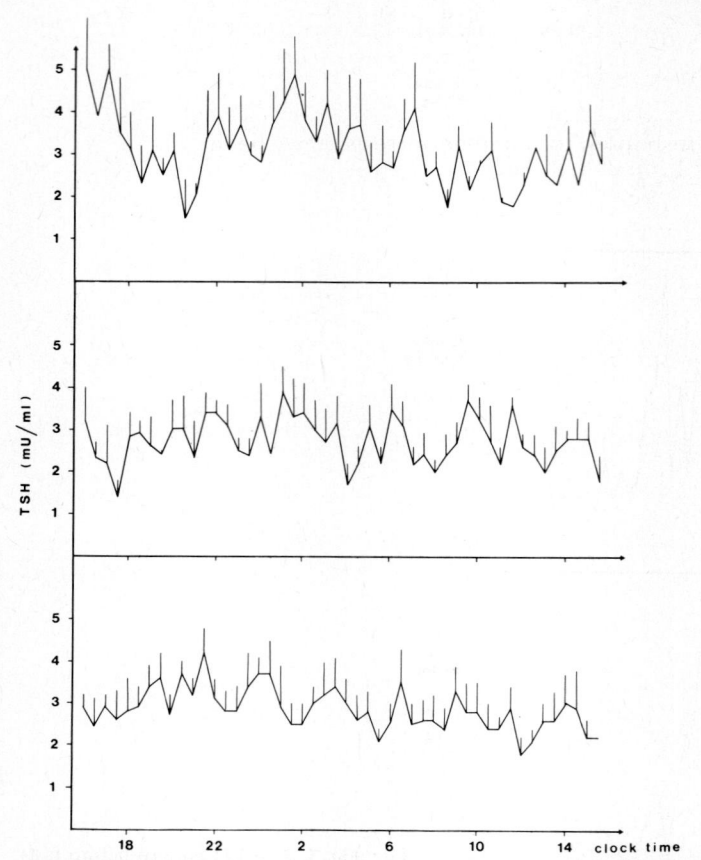

Fig. 4. 24-h TSH secretory pattern before fasting, after fasting, and after weight gain (mean and standard error)

In anorexia nervosa resting levels of thyrotropin (TSH) have been described as within normal limits in most cases. Thyrotropin responses to thyrotropin releasing hormone (TRH) have been reported as of normal magnitude by some authors, whereas others have found a blunted and delayed TSH response which may persist even after body weight has been restored (Casper and Frohman 1982). Our own results with normal healthy volunteers indicate that the pituitary thyroid function is less severely affected by starvation than the other hormonal axes. A diminished TSH response to stimulation with TRH was found in healthy subjects during starvation. Since the results concerning basal TSH levels and TSH following administration of TRH in anorexia nervosa are contradictory to some extent, further studies should provide clarification on this issue. Furthermore, it does not seem unlikely that reduced calorie intake or weight loss may also be one of the factors contributing to blunted TSH responses to TRH, which have been reported in depressed patients.

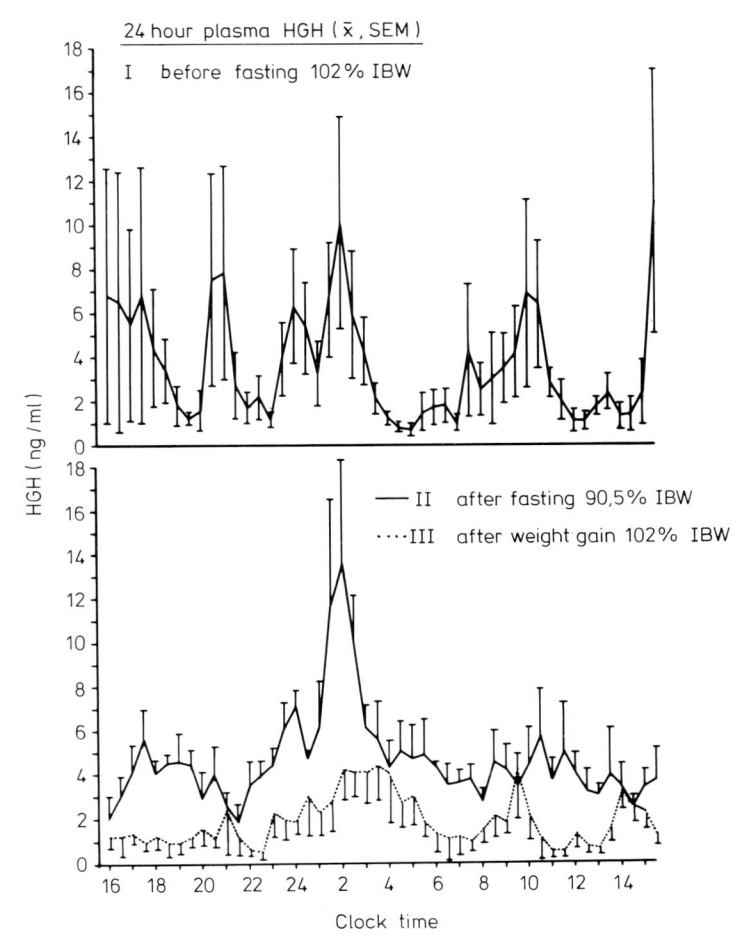

Fig. 5. 24-h growth hormone secretion before fasting, after fasting, and after weight gain (mean and standard error, in five subjects)

Growth Hormone (HGH)

The 24-h plasma growth hormone level (mean and standard deviation) showed an increase from 3.7 ± 2.6 ng/ml at the initial baseline to 4.8 ± 2.4 ng/ml after starvation in our healthy subjects (d.f. = 4, not statistically significant). The difference was more pronounced when the HGH level after fasting was compared with that after recovery to normal weight (2.2 ± 1.9 ng/ml; d.f. = 4, $P < 0.05$). Figure 5 shows the 24-h plasma HGH secretion before fasting, after fasting, and after weight gain. The sleep-related HGH peak after midnight was more pronounced after fasting. There appears to be a rebound phenomenon after recovery to normal weight: not only was HGH secretion lower then than after fasting, but also considerably lower than before fasting. A rebound-like effect was also observed for cortisol, LH, and prolactin.

Although a number of investigators have reported that resting growth hormone levels are elevated in anorexia nervosa, this has not been confirmed by others. Garfinkel et al. (1975) reported that HGH plasma levels normalized with recovery and that a fall in HGH levels was not related to changes in body weight but rather to caloric intake. HGH responses to stimulation tests have been variable in anorexia nervosa (paradoxical rise following oral glucose, reduced HGH response to apomorphine, flattened response to L-dopa, attenuated response to insulin, normal response to argenine, and increased response to TRH). Reasons for some of the discrepancies and findings concerning HGH secretion in anorexia nervosa are probably due to the fact that patients have been studied in various stages of the illness, in differing states of calorie intake, and with differences in eating behaviors reflecting the heterogeneity of patients concerning abstaining and purging. Our results recorded in healthy subjects show that reduced food intake results in a reversible increase in HGH secretion. Application of the α-adrenergic receptor agonist clonidine resulted in a normal HGH response after fasting; however, a subnormal HGH response was observed after weight gain (d.f. = 2, t-test for paired observations, $P < 0.05$). TRH application resulted in an increase in HGH secretion during starvation (not statistically significant).

Prolactin

In our starvation experiment the plasma prolactin level (mean and standard deviation) was significantly lowered, from 339 ± 127 to 223 ± 60 µU/ml (d.f. = 4, $P < 0.05$). After recovery to normal weight, plasma prolactin levels increased again to levels of 390 ± 99 µU/ml; this level was higher than the baseline levels (for II vs III, d.f. = 4, $P < 0.05$). Figure 6 shows the mean 24-h prolactin levels before weight loss, after weight loss, and after weight gain. Results indicate that starvation resulted in (a) a general lowering of prolactin levels, and (b) most of all the absence of a sleep-induced rise in nocturnal prolactin levels. After weight gain the sleep-related rise in nocturnal prolactin was seen again. A rebound effect was observed with a "more than normal secretory pattern" after recovery to normal weight.

There are relatively few reports on prolactin secretion in anorexia nervosa, and normal resting prolactin levels have been reported by several authors (Beumont et al. 1974; Mecklenburg et al. 1974; Vigersky and Loriaux 1977). Wakeling et al. (1979) found no relationship between basal prolactin levels and body weight, gonadotropin levels, and estradiol. Results concerning the prolactin response to stimulation with TRH have been conflicting; in our healthy normal subjects the increment in prolactin after stimulation with TRH was normal and response was not delayed throughout the study. It has been reported that the basal prolactin secretion pattern in normal adults shows a rise shortly after the onset of nocturnal sleep, which is followed by a series of episodes of still more pronounced secretion; the peak values of plasma prolactin can be observed during the end of the sleep period (Weitzman 1976). Low nocturnal prolactin levels and the absence of any sleep-induced rise in prolactin in anorexia nervosa

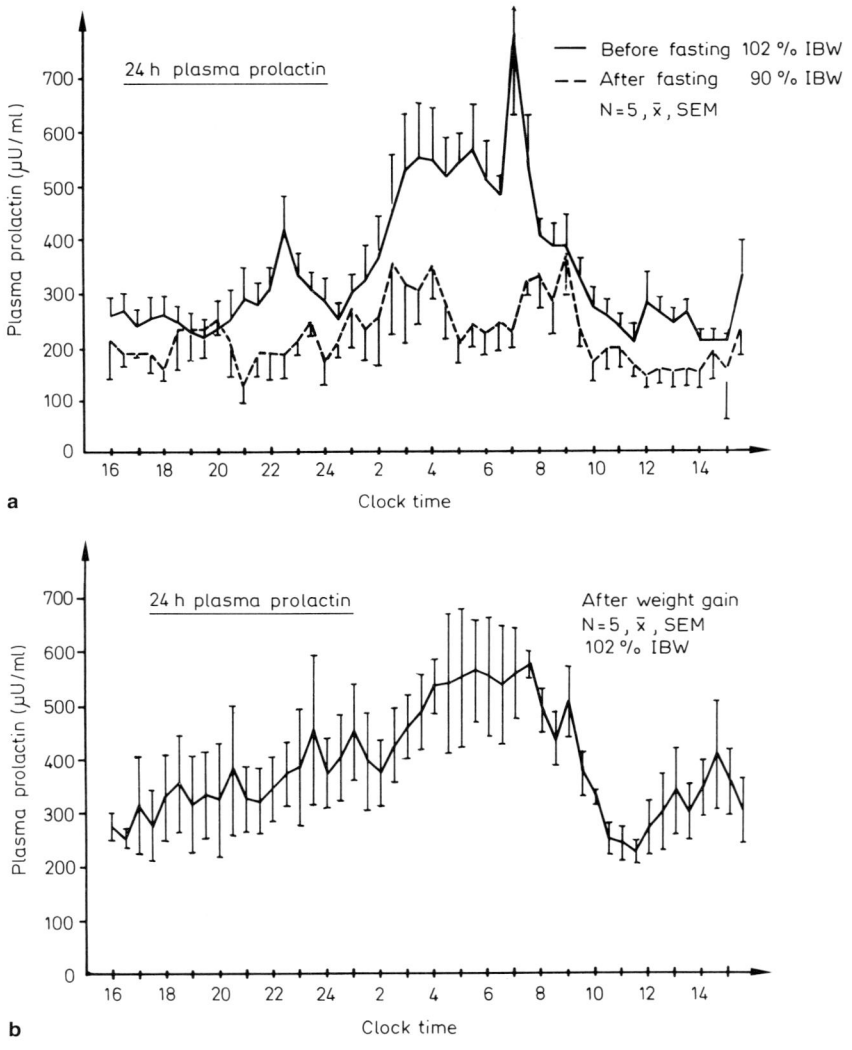

Fig. 6. 24-h plasma prolactin secretion pattern before fasting, after fasting, and after weight gain (mean and standard error, $n = 5$)

patients has been described by Kalucy et al. (1976), Brown et al. (1979), and Crisp (1980).

Our data in healthy starving subjects are in agreement with these latter observations. After fasting the healthy subjects showed a lowering of prolactin plasma levels at night and a reduction or absence of the sleep-induced prolactin peaks. In addition, the plasma prolactin levels were also lowered during daytime. These effects were reversible with weight gain. Considering these results it would be surprising to find normal prolactin secretion in all anorexia nervosa patients, for whom fasting is an essential feature. More refined and intensive longitudinal studies with larger series of patients will be needed to clarify the issue of lowered prolactin secretion in anorexia nervosa.

Conclusion

Hypothalamic pituitary function has been studied experimentally in healthy normal volunteers before and after complete starvation and after recovery to normal weight. Disturbances in endocrine function were detected in the hypothalamic—pituitary—adrenal axis, the LH secretory pattern, the TSH response to stimulation with TRH, basal growth hormone secretion, growth hormone response following stimulation with clonidine and basal prolactin secretion. The results support the hypothesis that the endocrine disturbances reported in anorexia nervosa are not primary but rather the consequence of reduced food intake or loss of body weight. Whether semistarvation has the same effects in healthy subjects as complete food abstinence, and which ingredients of nutrition are mainly responsible for the endocrine changes observed are questions open for empirical testing. Results of the present study are of relevance for the topic of starvation in general and with regard to anorexia nervosa. In addition the findings may well have implications for the interpretation of endocrine dysfunctions reported in other illnesses in which changes in weight and nutritional intake are common: for (endogenous) depression and alcoholism reduced food intake with weight loss appears to be at least one factor which significantly influences endocrine functions.

References

Beumont PJV, Russell J (1982) Anorexia nervosa. In: Beumont PJV, Burrows A (eds) Handbook of psychiatry and endocrinology. Elsevier, Amsterdam
Beumont PJV, Friesen HG, Gelder MG, Kolakowska T (1974) Plasma prolactin and luteinizing hormone levels in anorexia nervosa. Psychol Med 4: 219—221
Boyar RM, Katz J, Finkelstein JW, Kapen S, Weitzman ED, Hellman L (1974) Anorexia nervosa, immaturity of the 24-hour luteinizing hormone secretory pattern. N Engl J Med 291: 861—865
Boyar RM, Hellman LD, Roffwarg H, Katz J, Zumoff B, O'Connor J, Bradlow L, Fukishima DK (1977) Cortisol secretion and metabolism in anorexia nervosa. N Engl J Med 296: 190—193
Brown GM, Kirwan P, Garfinkel P, Moldofsky H (1979) Overnight patterning of prolactin and melatonin in anorexia nervosa (abstr). 2nd International symposium on clinical psycho-neuro-endocrinology in reproduction, Venice, June 1979
Casper RC, Frohman LA (1982) Delayed TSH release in anorexia nervosa following injection of thyrotropin-releasing hormone (TRH). Psychoneuroendocrinology 7: 59—68
Crisp AH (1980) Sleep, activity, nutrition and mood. Br J Psychiatry 137: 1—7
Doerr P, Fichter MM, Pirke KM, Lund R (1980) Relationship between weight gain and hypothalamic pituitary adrenal function in patients with anorexia nervosa. J Steroid Biochem 13: 529—537

Fichter MM, Doerr P, Pirke KM, Lund R (1982) Behavior, attitude, nutrition and endocrinology in anorexia nervosa. A longitudinal study in 24 patients. Acta Psychiatr Scand 66: 429–444

Garfinkel PE, Garner DM (1982) Anorexia nervosa. A multidimensional perspective. Brunner and Mazel, New York

Garfinkel PE, Brown GM, Stancer HC, Molodofsky H (1975) Hypothalamic pituitary function in anorexia nervosa. Arch Gen Psychiatry 32: 739–744

Kalucy RC, Crisp AH, Chard T, McNeilly A, Chen CN, Lacey JH (1976) Nocturnal hormonal profiles in massive obesity, anorexia nervosa and normal females. J Psychosom Res 20: 595–604

Mecklenburg RS, Loriaux DL, Thompson RH, Andersen AE, Lipsett MB (1974) Hypothalamic dysfunction in patients with anorexia nervosa. Medicine 53: 147–159

Pirke KM, Fichter MM, Lund R, Doerr P (1979) Twenty-four hour sleep-wake pattern of plasma LH in patients with anorexia nervosa. Acta Endocrinol (Kbh) 92: 193–204

Vigersky RA (1977) Anorexia nervosa. Raven, New York

Vigersky RA, Loriaux DL (1977) Anorexia nervosa as a model of hypothalamic dysfunction. In: Vigersky RA (ed) Anorexia nervosa. Raven, New York, pp 109–122

Walsh BT, Katz JL, Levin J, Kream J, Fukushima DK, Weiner H, Zumoff B (1981) The production rate of cortisol declines during recovery from anorexia nervosa. J Clin Endocrinol Metabol 53: 203–205

Wakeling A, DeSouza VA, Gore MBR, Sabur M, Kingston D, Boss AMB (1979) Amenorrhea, body weight and serum hormone concentrations, with particular reference to prolactine and thyroid hormones in anorexia nervosa. Psychol Med 9: 265–272

Weitzman ED (1976) Circadian rhythms and episodic hormone secretion. Annu Rev Med 27: 225–243

Perceptions of the Body in Anorexia Nervosa

P. E. Garfinkel and D. M. Garner[1]

The clinical features of anorexia nervosa are fairly similar, whether one reads descriptions from Morton (1694) or Gull (1868) or one sees patients today. Because of this clinical similarity, people have erred in looking for a single pathogenesis to the illness. In the sense of shared symptomatology, anorexia nervosa is clearly a discrete psychiatric syndrome, but as for other syndromes this does not imply a single pathogenesis. Rather, in any population there will be a group of individuals at risk for anorexia nervosa because of the presence of a specific combination of predisposing factors. It is the interaction and timing of these phenomena within a given individual which are necessary for the person to become ill. In this sense, anorexia nervosa is a final common pathway, the product of a group of interacting forces.

We have previously described a variety of factors within the individual, the family, and culture which have been purported to play a role in the pathogenesis or perpetuation of anorexia nervosa (Garfinkel and Garner 1982). One such area within the individual, that of body perceptions, will be examined in this paper by reviewing what is known of two major components of this: body image and interoception.

Body Image

Body image is a complex construct which has been approached from many points of view (Garner and Garfinkel 1981). These include the neural representation determining bodily experience (Head 1920), the mental image that one has of one's body (Traub and Orbach 1964), the feelings one has about one's body (Secord and Jourard 1953), and a personality construct (Fisher and Cleveland 1958; Kolb 1975; Schilder 1935). Reference to a disturbance in body image in anorexia nervosa dates back to Lasegue (1873): "The patient, when told that she cannot live upon an amount of food that would not support a young infant, replies that it furnishes sufficient nourishment for her, *adding that she is neither changed nor thinner.*" Bruch (1962), however, was the first to recognize disturbed body image as characteristic of anorexia nervosa. While she considers it to be related to a more general misperception of internal states, specifically it involves the patient's inability to recognize her appearance as abnormal.

---

1 Toronto General Hospital and University of Toronto, 200 Elizabeth Street, Toronto, Ontario, Canada M5G 1L7

If a broad definition of the body image construct is used, the disturbance in anorexia nervosa can be found to have different forms of expression, which may operate independently or conjointly. The first is "perceptual" and refers to the degree to which the patient is not able to assess her size accurately. The second involves cognitive and affective components without any obvious signs of "perceptual" mediators, and applies to some patients who assess their physical dimensions accurately but react to their bodies with extreme forms of disparagment or occasionally aggrandizement.

The most perplexing, yet most commonly described abnormality is the patient's apparent inability to recognize how thin she has become. Bruch (1973) refers to this as "disturbed size awareness"; the patient simply does not recognize that she has become emaciated. Some patients display a variation of this phenomenon in which the overestimation seems to be restricted to a particular part or parts of the body. Selected areas such as the stomach or thighs are magnified and seen as disproportionate to the rest of the body. These patients will acknowledge that in general they appear emaciated, but that further dieting is necessary to eliminate their "protruding stomach". Body image disturbance may be manifest in other patients who perceive their sizes relatively accurately but who exhibit an extraordinary loathing for all or parts of their body. This goes well beyond the dissatisfaction with their appearance common for Western women (Berscheid et al. 1973), to the point of revulsion with one's shape. A further expression of body image disturbance may occur either with normal size perception or with overestimation. This involves an exaggerated pleasure with, and overvaluation of, a thinner shape. These patients see their low weight and thinness as an exceptional achievement. Later this pleasure in low weight is replaced by a fear of any gain in weight.

Body Size Estimates as an Index of Body Image

A variety of techniques have been used to assess body image objectively; these have recently been described elsewhere (Garner and Garfinkel 1981; Garfinkel and Garner 1982) and will not be detailed here. In general, disturbed body image in anorexia nervosa has been defined as an overestimation of body parts or body size. We will confine this discussion to the latter and review work that has been done using a distorting photograph technique.

Glucksman and Hirsch (1969) had obese and normal subjects estimate their body sizes using a projected photograph which could be distorted along the horizontal axis. The image could be made to look anywhere from 20% "thinner" to 20% "fatter" than its actual size. Glucksman and Hirsch (1969) initially found that six dieting obese subjects overestimated their body sizes in comparison with four controls. Garner et al. (1976) applied this technique to the study of body size perception in anorectic and obese patients and found overestimation tendencies in many anorectics. Further, this overestimation of the body size was confined to the patient herself and did not include her estimates of inanimate objects or other people. More recently, Garfinkel et al. (1983b) have assessed body size estimates in the parents of anorectics, the anorectics themselves, and

Table 1. Results of body size estimation tests in parents and children

	Anorectics ($n = 23$) mean ± SD	Controls ($n = 12$) mean ± SD	Level of statistical significance
Child			
Visual self-perception body image (% over- or underestimation)	+ 8.1 ± 8.1	+ 1.3 ± 5.1	$P < 0.004$
Ideal image of self (% over- or underestimation)	− 5.4 ± 12.8	− 7.5 ± 6.9	n.s.
Mother			
Visual self-perception	+ 4.9 ± 7.2	+ 6.9 ± 7.4	n.s.
Ideal image of self	− 2.4 ± 8.7	− 3.4 ± 10.3	n.s.
Mother's estimate of child's size	+ 1.3 ± 7.6	+ 6.2 ± 6.9	n.s.
Mother's estimate of ideal child's size	+ 9.7 ± 8.4	+ 1.6 ± 6.1	$P < 0.004$
Father			
Visual self-perception	+ 7.4 ± 7.3	+ 6.5 ± 5.6	n.s.
Ideal image of self	+ 4.9 ± 9.9	+ 1.3 ± 9.7	n.s.
Father's estimate of child's size	− 2.3 ± 7.4	+ 5.5 ± 4.9	$P < 0.003$
Father's estimate of child's ideal size	+15.2 ± 5.1	+ 3.6 ± 7.1	$P < 0.001$

comparison nonanorectic families (Table 1). Overestimation occurred only in the anorectic patients and not in their parents. Moreover, parents of anorectics, quite appropriately, wished their daughters to be significantly larger (see Table 1). Mothers of anorectics were quite accurate in their assessment of their daughters' sizes, but fathers of anorectics underestimated their daughters' sizes relative to controls, probably because of their concerns regarding the illness. Garfinkel et al. (1983a) have also described a group of women with conversion disorders and serious weight loss; these individuals had lost large amounts of weight because of chronic vomiting. However, they did not display any of the cardinal features of anorexia nervosa. Vomiting was not due to a drive for thinness but rather usually expressed intense emotion, for example disgust over conflicts relating to sexuality or other areas. In comparison with a matched group of anorectics, the conversion group was found to lack body image disturbance (Table 2), suggesting that such self-overestimates are not merely due to weight loss or vomiting. Further evidence suggesting that body size overestimation is not a function of degree of weight loss comes from a recent study of Garner et al. (1983b). They studied women who had bulimia but who had never been emaciated. These "normal weight" bulimic subjects had body size distortions which were comparable to those in a bulimic group of anorectics. Other studies, using body part estimates of body image, have found less specificity to the overestimation phenomenon; using other techniques overestimation has been reported in many populations other than anorectics (Hsu 1982). There are different possible explanations for these discrepant findings: (a) The distorting photograph technique which measures general body size estimates may measure something more fundamental to anorectic's psychopathology than do techniques which assess body parts; or (b) overestimation may relate to particular psychopathologic phenomena which may be more

Table 2. Results of self-estimates of body size and ideal size in anorexia nervosa and conversion disorder with weight loss

	Conversion disorder ($n = 20$) mean ± SD	Anorexia nervosa ($n = 20$) mean ± SD	Level of statistical significance
Visual self-perception body image (% over- or underestimation)	− 3.2 ± 8.3	+4.8 ± 7.8	$P < 0.05$
Ideal image of self (% over- or underestimation)	+ 8.7 ± 17.9	+4.8 ± 11.7	$P < 0.05$

common in anorectics, but are not unique to them. This latter explanation will be discussed further, in terms of correlates of marked overestimation of body size.

While body size overestimation may not be specific to anorexia nervosa and while not all anorectics overestimate their sizes, anorectics as a group overestimate their sizes relative to normal women on the distorting photograph technique. Moreover, studies using both this technique and estimates of body parts have found overestimation to be a strong predictor of poor outcome (Garfinkel et al. 1977; Casper et al. 1979). The reliability of this technique has been demonstrated in studies that assessed the same patients at 1-week and 1-year intervals (Garfinkel et al. 1978, 1979). It has also been found to be stable in spite of such manipulations as having subjects study their images in a mirror or ingest meals that connote high- or low-calorie foods (Garfinkel et al. 1978).

"Marked" Overestimation of Body Size

Our group has studied 265 consecutive anorectic patients (using the criteria of Garfinkel and Garner 1982) with this technique and with an associated battery of psychometric tests. We have found that anorectics as a group overestimate their body size by over 4%, significantly more than control women of similar age, who as a group are quite accurate (Fig. 1). Moreover, a large subgroup (40%) overestimate their body size by more than 10%, a degree of overestimation which is extremely uncommon in normal women (Garner et al. 1976; Garfinkel et al. 1978). When we compared these 105 "marked overestimators" with the 160 anorectic patients who overestimate only moderately or underestimate their body size, we found that the marked overestimators weighed significantly less than the others (73.4% vs 79.7% of average, $P < 0.001$). Because this could have a significant effect on other variables we selected subgroups of these of 88 subjects each who were matched for weight. We have compared these two groups on a number of clinical, historical, and psychometric variables in order to learn more about this group who markedly overestimate their sizes.

Table 3 displays data relating to the two groups on a variety of clinical phenomena relating to weight and duration of illness. While the marked

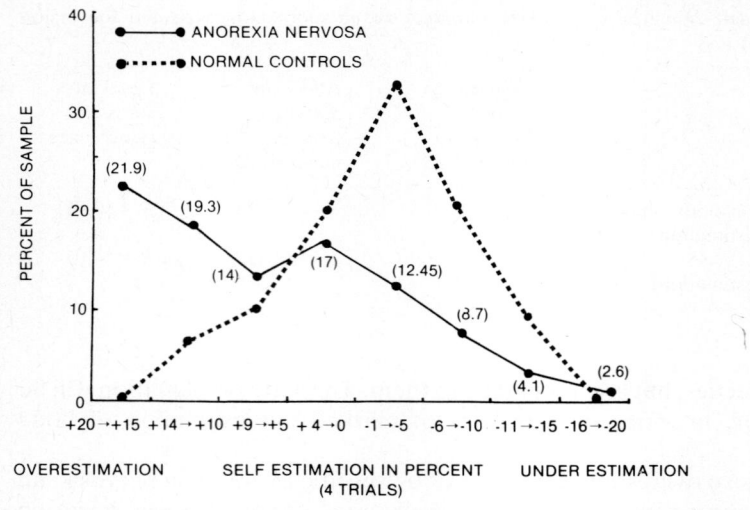

Fig. 1. Self-estimation of body size in anorectics ($n = 265$) and normal controls

Table 3. Comparison of marked overestimators and other anorectics on clinical and historical information

	Marked overestimators ($n = 88$) mean ± SD	Underestimators and moderate overestimators ($n = 88$) mean ± SD	Level of statistical significance
Weight on testing (kg)	46.9 ± 9.8	45.9 ± 8.8	n.s.
Percentage of average weight	78.6 ± 13.5	78.8 ± 13.9	n.s.
Maximum weight (kg)	61.1 ± 11.8	60.5 ± 10.3	n.s.
Percentage of average at maximum	102.7 ± 17.6	103.1 ± 15.2	n.s.
Minimum weight (kg)	40.4 ± 6.9	41.1 ± 6.9	n.s.
Percentage of average at minimum	67.9 ± 9.6	70.1 ± 9.9	n.s.
Age (years)	22.1 ± 5.6	21.7 ± 5.1	n.s.
Duration of illness (months)	49.3 ± 54.9	41.2 ± 41.1	n.s.
Social class	2.8 ± 1.7	2.6 ± 1.5	n.s.

overestimators tended to be individuals with bulimia, vomiting, and laxative misuse, these differences were not statistically significant. Marked overestimators more commonly did not eat any breakfast ($P < 0.05$), lunch ($P < 0.01$), and dinner ($P < 0.05$). Important differences between the groups related to depressive symptoms; the marked overestimators were described as having labile moods ($P < 0.05$), fragmented sleep ($P < 0.05$), and early morning awakening ($P < 0.01$). They had also mutilated themselves more commonly than the comparison group (18% vs 7%, $P < 0.05$).

Table 4 displays results comparing the two groups on psychometric measures of attitudes toward eating, weight, and their bodies. On each of the measures of

Table 4. Comparison of marked overestimators and other anorectics on attitudes to eating, weight, and their bodies

	Marked overestimators mean ± SD	Underestimators and moderate overestimators mean ± SD	Level of statistical significance
Eating attitudes test (EAT) (Garner and Garfinkel 1979)	45.3 ± 14.7	33.3 ± 15.6	$P < 0.001$
Restraint (Herman and Polivy 1975)	25.2 ± 5.1	22.4 ± 6.2	$P < 0.002$
Body satisfaction (Berscheid et al. 1973)	80.4 ± 18.3	69.6 ± 14.6	$P < 0.001$
Anhedonia (Chapman et al. 1976)	17.0 ± 9.1	13.1 ± 7.0	$P < 0.004$
Ideal body size estimate (Garner et al. 1876)	− 6.1 ± 12.6	− 0.1 ± 13.5	$P < 0.003$

Table 5. Comparison of marked overestimators and other anorectics on other psychometric tests

	Marked overestimators mean ± SD	Underestimators and moderate overestimators mean ± SD	Level of statistical significance
Beck Depression Inventory (Beck 1978)	30.4 ± 13.9	21.0 ± 12.5	$P < 0.001$
Hopkins Symptom Checklist (Derogatis et al. 1974) (total)	136.9 ± 32.0	116.8 ± 29.4	$P < 0.001$
Somatization	23.5 ± 7.0	20.0 ± 6.1	$P < 0.001$
Anxiety	17.0 ± 5.3	14.1 ± 6.1	$P < 0.001$
Depression	29.1 ± 7.2	25.0 ± 7.6	$P < 0.001$
Obsessive-compulsive	18.9 ± 6.1	15.8 ± 5.4	$P < 0.002$
Interpersonal sensitivity	17.4 ± 4.3	15.5 ± 4.5	$P < 0.01$
Locus of control (Reid and Ware 1973)	18.6 ± 6.6	16.0 ± 5.8	$P < 0.01$
Test for self-esteem (Janis and Field 1959)	72.6 ± 14.5	66.2 ± 12.4	$P < 0.05$

attitudes toward eating and weight and dieting (EAT and Restraint) the marked overestimators displayed more extreme scores. Furthermore, on the affective component of body image, body dissatisfaction, the overestimators again revealed greater displeasure with their bodies; they also reported significantly more anhedonia. Corresponding with these findings, the marked overestimators reported wishing to be significantly thinner on the distorting photograph technique, in spite of actually weighing the same as the other anorectics.

Table 5 shows results of the marked overestimators on other psychometric tests. Their increased depression is evident on the Beck Depression Inventory, as well as on the depression subscale of the Hopkins Symptom Checklist. They also scored as more externally controlled, with lower self-esteem (Janis-Field 1959) and greater psychopathology in general. Therefore marked overestimation is both a poor prognostic sign (Garfinkel et al. 1977) and also associated with psychopathology on a number of parameters.

A number of different theories have been offered as explanations for body image disturbance in anorexia nervosa; these have been reviewed elsewhere (Garner and Garfinkel 1981). Of importance here, the data on the interrelationships between marked overestimation of size, body dissatisfaction, anhedonia, depression, and altered self-esteem suggest that body perceptual abnormalities are closely related to other dimensions of negative self-concept. This explanation may help explain why overestimation of size is common in, but not unique to, anorexia nervosa. This is in agreement with self-esteem consistency theory (Wylie 1968), which holds that expectations and perceptions will be determined by one's evaluation of general, nonphysical attributes. Further evidence for this comes from work comparing women with anorexia nervosa with very weight-preoccupied nonanorectic women (Garner et al., to be published a). This latter group were selected because they had a desire for thinness equal to the anorectics but they never developed the clinical syndrome of anorexia nervosa. These weight-preoccupied women displayed dissatisfaction with their bodies equal to that of the anorectics and far greater than in normal women. However, they lacked other psychological parameters of anorectics. Therefore a strong drive for thinness may be associated with a general dissatisfaction with one's body and negative sense of self, and this occurs in both anorectic and very weight-preoccupied nonanorectic women. Further studies on normal-weight women with bulimia (Garner et al., to be published b) also support a link between body image disturbance and these measures of psychopathology.

It has been asserted that in anorexia nervosa, self-worth becomes concretized onto body shape (Bruch 1973). Anorectics become highly sensitized to shape and body fatness becomes an index by which nonphysical qualities are evaluated. If an individual views all aspects of herself negatively, as anorectics with a dichotomous ("all or nothing") thinking style are prone to do (Garner and Bemis 1982), and if she also equates low self-esteem with fatness, she may see herself as larger than her actual size. Low self-esteem and body dissatisfaction may play a role in misperception of size in anorexia nervosa, particularly when coupled with heightened shape concerns.

Distortions in Internal Perceptions

According to Bruch (1962, 1973) the concept of "body awareness" or "body identity" is not limited to body image but extends to the perception and interpretation of interoceptive stimuli. More specifically, she postulates that anorexia nervosa is fundamentally related to disturbed awareness of inner processes which include misperception of hunger, satiety, and other bodily sensations. She considers the lack of responsiveness to fatigue, cold, and sexual feelings in anorexia nervosa to be examples of this disturbance.

Patients will often describe extreme confusion about their internal states early in treatment or will appear devoid of thoughts and feelings which reflect their personal experiences. These deficits range in depth from subtle confusion in affective labeling to complete mistrust of one's internal state. While variable in severity, this disturbance is very common in anorexia nervosa [for further clinical descriptions of this see Bruch (1973) and Garfinkel and Garner (1982)].

Consistent with the clinical observations of interoceptive disturbances, several lines of experimental inquiry suggest that these patients may not experience their internal environment in the same way as individuals without eating disorders.

In the early part of this century, Cannon (1912) claimed that gastric contractions, measured by an intragastric balloon, could be related to hunger sensations in normal individuals. The relationship between gastric activity and "hunger" has not been supported by more recent investigations (Penick et al. 1967; Stunkard and Fox 1971). However, the perception of gastric contractions in anorexia nervosa has been examined by Silverstone and Russell (1967) using an intragastric tube, and by Crisp (1967) using an intragastric pressure telemetry pill. Both studies found no significant differences in stomach motility between anorectic patients and normal subjects. The anorectic patients were capable of recognizing contractions, but interestingly, some did not interpret these as sensations of hunger.

Coddington and Bruch (1970) found that both anorectic and obese subjects were less accurate than normals in perceiving the amounts of food (Metrecal) that were directly introduced into their stomachs. While only three anorectic subjects were included in this study, it does support the hypothesized deficit in the recognition of internal state. Other studies have reported that subjects with juvenile onset obesity have difficulties appropriately responding to internal satiety cues (Cabanac and Duclaux 1970; Cambell et al. 1971) or that their eating behavior may be largely determined by external circumstances such as the availability, salience, and palatability of food (Nisbett 1972; Schachter 1971). However, this external responsiveness may be a phenomenon of dieting, rather than obesity (Herman and Polivy 1975), and therefore probably applies to anorectics as well as the obese.

The perception of hunger and satiety in anorexia nervosa was investigated by Garfinkel (1974) using several self-report questionnaires. Eleven anorectic patients reported distorted sensations of satiety compared with II control subjects. After fasting for 12 h, all subjects were given a questionnaire inquiring about the experience of hunger. Subjects were then given a standard meal followed by another questionnaire assessing satiety. Except for an increased preoccupation with food and a fear of eating, anorectics did not differ from normals in their perception of hunger. However, in contrast to normals, the anorectic patients reported disturbed sensations of satiety including bloating, absence of stomach sensations, nausea, and aches and pains. In a recent study, Dubois et al. (1979) have shown that anorectic patients display delayed gastric emptying compared with normal controls. After weight gain, the emptying rates for the anorectic patients tended to increase toward normal but were still significantly less than in controls. These results may partially explain the postprandial fullness, discomfort, and early satiety observed with most patients.

In a recent series of studies Garfinkel et al. (1978, 1979) have used taste perception as an index of satiety in anorexia nervosa. This procedure was derived from the work of Cabanac and Duclaux (1970), who found that obese subjects do not experience any differences in the rated pleasantness of sucrose tastes before versus after the ingestion of glucose. Normal subjects in contrast, experience "satiety" or an aversion to the taste of sucrose after glucose preloading. On the basis of these results Cabanac and Duclaux (1970) suggested that obese subjects are less responsive than nonobese to internal cues related to nutritional requirements. A replication of the Cabanac and Duclaux study (Underwood et al. 1973) reported an interesting but unanticipated finding. One of the normal subjects who displayed an absence of aversion to sucrose was later found to have anorexia nervosa.

Our first study of this demonstrated that disturbances in interoception measured by the satiety-aversion-to-sucrose test were evident in anorectic patients in contrast to normal controls. A modified version of the Cabanac and Duclaux (1970) procedure was adopted; only a 20% sucrose solution was used, since this produced maximal differences between obese and control subjects in the Cabanac and Duclaux (1970) studies. Rather than using a 50-g glucose load, we used two test lunches which contained 400 kcal. One of these was designed to connote a high calorie content; it consisted of tuna salad, coleslaw, and a large chocolate sundae. The second lunch was identical in acutal calorie content but was designed to connote a relatively low calorie content; it consisted of tuna salad and coleslaw both supplemented with gluconal. Sweetness and pleasantness ratings were collected on all subjects every 3 min for 1 h after lunch. The anorectics, unlike the controls, failed to develop an aversion to the sweet taste after the test lunches. Follow-up of many of these patients 1 year later revealed similar results.

Furthermore, in a large sample of our anorectic patients this interoceptive disturbance was related to the tendency to overestimate body size ($n = 72, r = 0.26, P < 0.02$); it is the overestimators who fail to develop a normal aversion. In contrast to Crisp and Kalucy's (1974) earlier report, we did not find that the caloric connotation of the meal systematically influenced body size estimation in either patients or controls. The lack of satiety aversion was also independent of which meal was eaten. As in an earlier study (Garfinkel 1974) the results on a satiety questionnaire differentiated anorectics from normal subjects on various parameters. These were postprandial changes in mood, gastrointestinal sensations, and willpower required to stop eating. Using a 10-cm analog scale for satiety, anorectics displayed greater premeal fullness and their fullness persisted longer after eating the meal that they believed to have more calories. This latter finding suggests a substantial cognitive influence in the satiety experience in anorectic patients.

While there is evidence for a cognitive component to the altered satiety experience, recent work also suggests that the changes in gastrointestinal physiology that accompany starvation may also play a role. Starving anorectics have delayed gastric emptying (Dubois et al. 1979; Saleh and Lebwohl 1979; Holt et al. 1981), which probably contributes to many symptoms of satiety following a meal. Dubois et al. (1979) showed that bethanacol, a cholinergic agonist, partially improved gastric emptying, while Saleh and Lebwohl (1980) demonstrated that metoclopromide returned gastric emptying to normal. However, the use of metoclopromide is limited by neurological side-effects and its propensity to produce depression. Recently Russell et al. (1983) have described an anorectic patient with delayed gastric emptying and subjective postprandial bloating. Domperidone, a novel compound which enhances gastric peristalsis, accelerates gastric emptying time, and does not cross the blood-brain barrier, improved both gastric emptying and the patient's satiety ratings. A controlled trial of this medication is now in progress.

In summary, several lines of experimental inquiry have suggested that patients with anorexia nervosa may misperceive internal experiences. It is not clear whether these perceptual disturbances are determinants or by-products of the syndrome, and what their relationship is to complete recovery. Since these aberrant experiences are clinically important with these patients, more detailed study of potential mechanism will be of considerable value.

References

Beck AT (1978) Depression inventory. Center for Cognitive Therapy, Philadelphia
Berscheid E, Walster E, Hohrnstedt G (1973) The happy American body: a survey report. Psychology Today November: 119–131
Bruch H (1962) Perceptual and conceptual disturbances in anorexia nervosa. Psychol Med 24: 187–194
Bruch H (1973) Eating disorders: obesity, anorexia nervosa and the person within. Basic Books, New York
Cabanac M, Duclaux R (1970) Obesity: absence of satiety aversion to sucrose. Science 168: 496–497
Campbell RG, Hashim SA, VanItallie TB (1971) Studies of food intake regulation in man: responses to variations in nutritive density in lean and obese subjects. N Engl J Med 285: 1402–1407
Cannon WB, Washburn AL (1912) An explanation of hunger. Am J Physiol 29: 441–454
Casper RC, Halmi KA, Goldberg SC, Eckert ED, Davis JM (1979) Disturbances in body image estimation as related to other characteristics and outcome in anorexia nervosa. Br J Psychiatry 134: 60–66
Chapman LJ, Chapman JP, Raulin ML (1976) Scales for physical and social anhedonia. J Abnorm Psychol 85: 374–382
Coddington RD, Bruch H (1970) Gastric perceptivity in normal, obese and schizophrenic subjects. Psychosomatics 11: 571–579
Crisp AH (1967) The possible significance of some behavioural correlates of weight and carbohydrate intake. J Psychosom Res 11: 117–131
Crisp AH, Kalucy RS (1974) Aspects of the perceptual disorder in anorexia nervosa. Br J Med Psychol 47: 349–361
Derogatis L, Lipman R, Rickels K, Uhlenhuth EH, Covi L (1974) The Hopkins symptom checklist (HSCL): a self report symptom inventory. Behav Sci 19: 1–15
Dubois A, Gross HA, Ebert MH, Castell DO (1979) Altered gastric emptying and secretion in primary anorexia nervosa. Gastroenterology 77: 319–323

Fisher S, Cleveland SE (1958) Body image and personality. Dover, New York
Garfinkel PE (1974) Perception of hunger and satiety in anorexia nervosa. Psychol Med 4: 309–315
Garfinkel PE, Garner DM (1982) Anorexia nervosa: a multidimensional perspective. Brunner Mazel, New York
Garfinkel PE, Moldofsky H, Garner DM (1977) Prognosis in anorexia nervosa as influenced by clinical features, treatment and selfperception. Can Med Assoc J 117: 1041–1045
Garfinkel PE, Moldofsky H, Garner DM, Stancer HC, Coscina DV (1978) Body awareness in anorexia: disturbances in body image and satiety. Psychosom Med 40: 487–498
Garfinkel PE, Moldofsky H, Garner DM (1979) The stability of perceptual disturbances in anorexia nervosa. Psychol Med 9: 703–708
Garfinkel PE, Kaplan AS, Garner DM, Darby PL (1983a) The differentiation of vomiting and weight loss as a conversion disorder from anorexia nervosa. Am J Psychiatry 140: 1019–1022
Garfinkel PE, Garner DM, Rose J, Darby PL, Brandes JS, O'Hanlon J, Walsh N (1983b) A comparison of characteristics in the families of patients with anorexia nervosa and normal controls. Psychol Med 13: 821–828
Garner DM, Bemis K (1982) A cognitive-behavioral approach to anorexia nervosa. Cognitive Ther Res 6: 1–27
Garner DM, Garfinkel PE (1981) Body image in anorexia nervosa: Measurement, theory and clinical implications. Int J Psychiatry Med 11: 263–284
Garner DM, Garfinkel PE, Stancer HC, Moldofsky H (1976) Body image disturbances in anorexia nervosa and obesity. Psychosom Med 38: 227–236
Garner DM, Olmsted M, Garfinkel PE (to be published a) A comparison between weight preoccupied women and anorexia nervosa. Int J Eating Disorders
Garner DM, Garfinkel PE, O'Shaughnessy M (to be published b) A comparison between bulimia with and without anorexia nervosa. New directions in anorexia nervosa. Report of the fourth ross conference on medical research
Glucksman ML, Hirsch J (1969) The response of obese patients to weight reduction. Psychosom Med 31: 1–7
Gull WW (1868) The address in medicine delivered before the annual meeting of the BMA at Oxford. Lancet 2: 171
Head H (1920) Studies in neurology. Hodder Stoughton, London
Hermann CP, Polivy J (1975) Anxiety, restraint and eating behaviour. J Pers 84: 666–672
Holt S, Ford MJ, Grant S (1981) Abnormal gastric emptying in primary anorexia nervosa. Br J Psychiatry 139: 550–552
Hsu LK (1982) Is there a disturbance in body image in anorexia nervosa. J Nerv Ment Dis 170: 305–307
Janis IL, Field PB (1959) Sex differences and personality factors related to persuasibility. In: Houland CI, Janis IL (eds) Personality and persuasibility. Yale University Press, New Haven, pp 55–68
Kolb C (1975) Disturbances of body image. In: Arieti (ed) American handbook of psychiatry. Basic Books, New York, pp 810–831
Lasègue (1873) De l'anorexie hystérique. Reprinted (1964) in: Kaufman RM, Heiman M (eds) Evolution of psychosomatic concepts. Anorexia nervosa: a paradigm. International Universities Press, New York
Morton R (1694) Phthisiologica: or a treatise of consumptions. Smith and Wolford, London
Nisbett RE (1972) Eating behaviour and obesity in men and animals. Adv Psychosom Med 7: 173–193
Penick SB, Smith GP, Wieneke K Jr, Hinkle LE Jr (1967) An experimental evaluation of the relationship between hunger and gastric motility. Am J Physiol 205: 421–426
Reid DW, Ware EE (1974) Multidimensionality of internal versus external control: addition of a third dimension and nondistinction of self versus others. Can J Behav Sci 6: 131–142
Russell D McR, Freedman ML, Feiglin DHI, Jeejeebhoy KN, Swinson RP, Garfinkel PE (1983) Delayed gastric emptying in anorexia nervosa – improvement with domperidone. Am J Psychiatry 140: 1235–1236

Saleh JW, Lebwohl P (1980) Metoclopramide-induced gastric emptying in patients with anorexia nervosa. Am J Gastroenterol 74: 127–132
Schachter S (1971) Emotions, obesity and crime. Academic Press, New York
Schilder P (1935) Image and appearance of the human body. Kegan, London
Secord PF, Jourard SM (1953) The appraisal of body cathexis: body cathexis and the self. Consult Psychol 17: 343–347
Silverstone JT, Russell GFM (1967) Gastric hunger in contractions in anorexia nervosa. Br J Psychiatry 13: 257–263
Stunkard AJ, Fox S (1971) The relationship of gastric motility and hunger: a summary of the evidence. Psychosom Med 33: 123–134
Traub AC, Orbach J (1964) Psychological studies of body image: an adjustable body distorting mirror. Arch Gen Psychiatry 11: 53–66
Underwood PJ, Belton E, Hulme P (1973) Aversion to sucrose in obesity. Proc Nutr Soc 32: 94
Wylie R (1968) The present status of self-theory. In: Borgatta EG, Lambart WW (eds) Handbook of personality theory and research. Rand McNally, Chicago

Treatment of Anorexia Nervosa: What Can Be the Role of Psychopharmacological Agents?

A. H. Crisp[1]

In approaching this problem, can we learn anything from a knowledge of the drugs and similar substances which people with anorexia nervosa sometimes consume in attempts to sustain their adjustments? It must be a moot point as to whether purgatives and diuretics can be psychopharmacological agents! However, anorectics certainly consume them because of the desirable psychological consequences associated with their dehydrating effects — the experiences and self-perceptions of reduced weight, sunken facies, flattened abdomen, etc. In the past a few anorectics have taken to consuming large amounts of amphetamine compounds and/or thyroid medication but, as well as being hard to come by, these drugs have never been popular with the majority of anorectics for reasons that I will touch on later. Some anorectics become dependent on hypnotics, including alcohol, and minor tranquilizers; others insist on taking the contraceptive pill. I believe that an inspection of this range of drugs tells us that the anorectic's task in life is to maintain her low body weight so that she feels reasonably secure. However, her panic at the prospect of losing control over her weight and her aroused and exhausted state mean that she is still generally anxious, tense, and weary. At the same time she wishes to be seen as normal and not in need of treatment. Against this background it is well to remember that suicide is the commonest cause of death in anorexia nervosa and that psychotropic drug overdose is the commonest form of suicidal act in this population. Most clinicians attempting to treat anorectics would consider it desirable for their patients to stop taking most of these drugs, but does such self-medication give us any clues as to whether there are other pharmacological agents that might be of value in attempts to help the person with anorexia nervosa shed her condition?

When I first became especially interested in anorexia nervosa around 1960, I and a colleague reported a case (Crisp and Roberts 1962) which has influenced my thinking on anorexia nervosa in several respects ever since (Fig. 1). This patient showed the profound metabolic changes characterizing the severely emaciated, abstaining anorectic (Crisp 1967). He displayed that paradox wherein the energy-deprived and energy-conserving anorectic is nevertheless highly aroused: fated to be endlessly preoccupied with thoughts of food, to forage and wrestle with the starving impulse to ingest, all in the face of overriding terror of weight gain and a desperate wish, in the presence of plenty of food, never to have to think about it again. I treated him with psychotherapy of a similar nature to the individual and family approach I still adopt today, and

---

1 University of London, Department of Psychiatry, St. George's Hospital Medical School, London SW 17 ORE, Great Britain

The Psychobiology of Anorexia Nervosa
Edited by K. M. Pirke and D. Ploog
© Springer-Verlag Berlin Heidelberg 1984

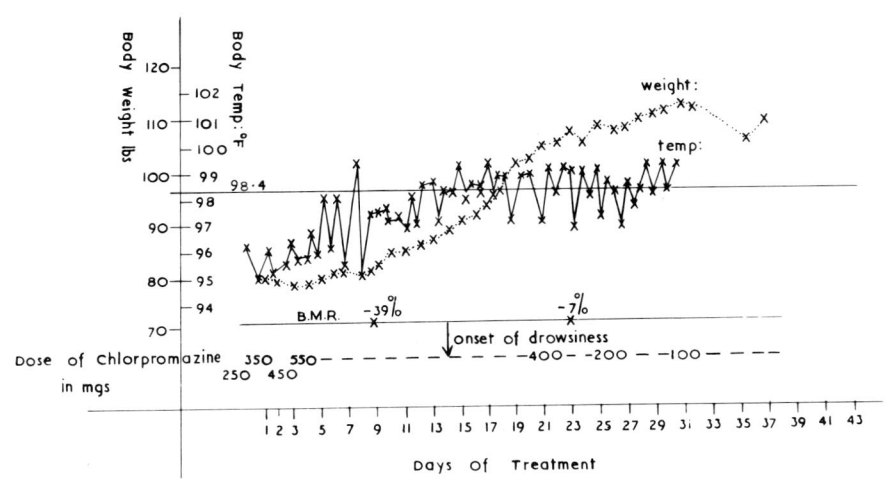

Fig. 1. As this young male anorectic gained weight there was a threshold change in his metabolism, reflected in his body temperature and BMR, which coincided with the onset of drowsiness in relation to his dosage of chlorpromazine (Crisp and Roberts 1962)

also with large doses of chlorpromazine. This treatment was fashionable at the time (Dally and Sargant 1960). It was a clinical view around that time that the ability to tolerate large doses of chlorpromazine without succumbing to its sedative effects was often indicative of psychosis. My patient was indeed highly aroused through starvation. As his weight increased there came a time when he changed rapidly in respect of his physical status. His resting metabolic rate increased step-like and his body temperature dramatically reflected this, reverting in the process to its normal diurnal rhythm instead of the converse. Coincidentally, he became drowsy and his chlorpromazine dosage was reduced. These changes coincided, in my view, with restoration of full biological function following the reactivation of the pubertal process associated with weight gain and with a significant move towards restoration of normal energy balance. In retrospect the chlorpromazine seemed to have had very little direct effect on his appetite or ingestion, although it had probably rendered him more compliant, muting his ambivalence about treatment, and less active, having conferred extrapyramidal immobility upon him. For some years after I continued to use chlorpromazine as such an aid to treatment (Crisp 1965).

These days I rarely use any drug treatment, although occasionally I find certain drugs useful as aids to treatment. Indeed we have treated hundreds of severely ill anorectics without drugs — with them regaining weight to fully adult proportions (mean weight on admission 38 kg; mean weight on discharge 54 kg) — through the process of first enabling them to become 'patients' and thereby contemplate real change (Crisp 1980b). However, there continue to be advocates for drug treatments, sometimes as the primary and major approach. The two kinds of drugs most commonly used under these circumstances are: (a) those known to be associated with weight gain in other settings, and (b) those of a supposedly specifically antidepressant kind, on the basis that anorexia nervosa has a strand or more than a strand of depressive illness within it. Although there is overlap between these two categories of drugs, I will describe them separately.

Drug Treatments Associated with Weight Gain in Psychiatric Patients

1. Antidepressants. The tricyclic and tetracyclic drugs. The propensity of these drugs, especially the sedative group, to generate weight gain independent of their effect on mood was highlighted by Paykel et al. (1973), who described carbohydrate craving as an effect of amitryptyline
2. Antipsychotic agents, e.g., the phenothiazines, thioxanthines, and butyrophenones (Amdisen 1964; Singh et al. 1970).
3. Mood-stabilizing drugs, e.g., lithium carbonate (Vendsborg et al. 1976).
4. Antiserotoninergic (and antihistaminic) drugs, e.g., cryproheptadine, pizotifen, and methysergide (Silverstone and Schuyler 1975).

The modes of action of these drugs in respect of weight gain are uncertain. Sometimes more than one contributing effect on weight gain can be identified. They include the following:

1. Increased hunger/ingestion (Paykel et al. 1973).
2. Increased fluid intake. This has been implicated, for example, as one of the reasons for weight gain associated with ingestion of chlorpromazine (see Crisp 1965). Other drugs have the well-known side-effect of producing a dry mouth and provoking excessive fluid intake through this mechanism. Occasionally, there can be a real alteration in fluid balance homeostasis; for example, it has been suggested that with the use of lithium carbonate a functional form of diabetes insipidus is created, which results in excess fluid loss. The patient drinks extra, and often heavily calorie-laden, fluids to compensate (Vendsborg et al. 1976). In seeking to identify such specific effects of drugs it must be remembered that increased food intake itself leads to some fluid retention.
3. Altered utilization. Once again chlorpromazine may have its effect partly through this mechanism. The drug reduces resting metabolism.
4. Increased availability of food, e.g., reduced opportunity to avoid eating food, throwing it away, or vomiting it. Clearly, the setting associated with specific treatments may affect this variable.
5. Decreased activity: the extrapyramidal and motor disorders associated with administration of some of these drugs can reduce overall activity and energy expenditure.
6. Increased compliance. It was probably the increased compliance of schizophrenics taking large doses of chlorpromazine that accounted for most of their weight gain in the early mental hospital studies of this phenomenon.
7. Premorbid obesity/history of weight fluctuation. These characteristics are associated with the greatest weight fluctuations in presenting psychiatric

illnesses (Crisp and Stonehill 1976). Treatment aimed at such illnesses is likely then to be associated with weight gain as a largely nonspecific effect.

Of these drugs, phenothiazines (as previously described) and antidepressants (see below) have been most widely used in attempts to generate weight gain in individuals with anorexia nervosa. Halmi et al. (1983) have reported on the use of amitriptyline and cyproheptadine. She compared these drugs with a placebo in a double blind trial involving a random assignment of patients to the three treatments. She claims a special effect of cyproheptadine in promoting weight gain against a standard background of hospital treatment. Weaknesses in the study include a placebo population which was idiosyncratic in terms of relatively limited weight gain prior to the treatment study and the fact that many of her criteria for identifying depression (she administered the drugs on the grounds that they are antidepressants) are in fact nonspecific factors (see below). At first glance it seems that her two patient groups who received the active drugs more or less achieved normal weight. In fact, the authors define normal weight as ideal body weight as indicated by Metropolitan Life Insurance scales. These weights, of course, will be significantly lower than average population weights. Earlier, Lacey and Crisp (1980) had demonstrated that clomipramine given under standard background treatment circumstances was associated with reports of increased appetite (Fig. 2) and hunger and with increased energy intake (Fig. 3) when the patient was allowed to select her own diet. This effect disappeared once the patient had achieved a pubertal-type body weight and was on the whole less pronounced than that seen within the alternative treatment setting where patients were expected to eat 3,000 calories per day as part of the treatment contract. However, clomipramine was also associated with greater physical activity than placebo. This may have been a direct effect of the drug (Crisp 1980a) or else defensive. On 3-year follow-up there were no differences between the clomipramine and placebo groups in this study.

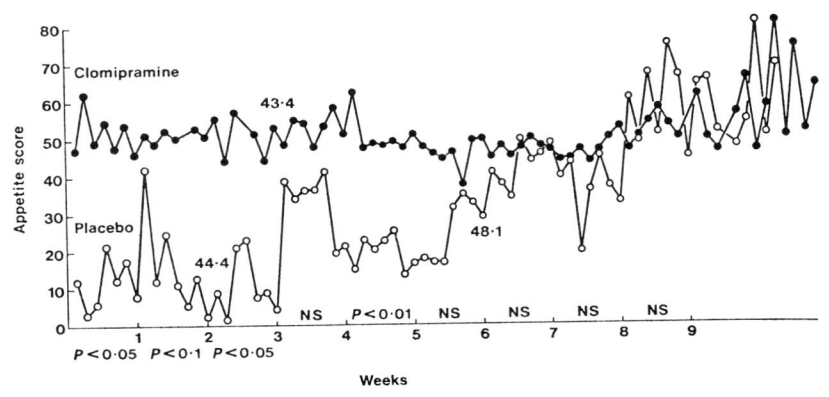

Fig. 2. Mean daily appetite scores (analog scale) of anorectic patients gaining weight and receiving clomipramine or placebo (Lacey and Crisp 1980)

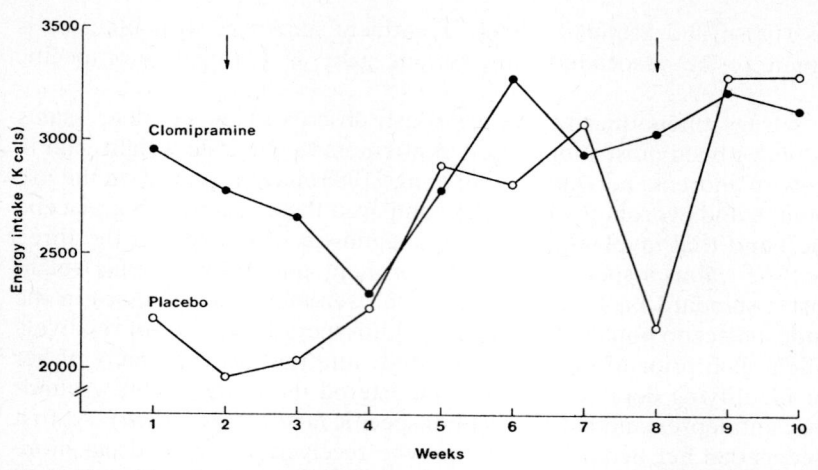

Fig. 3. Mean "free day" energy intake of two groups of anorectics gaining weight and concurrently receiving either clomipramine or placebo. There was one such free day within each week and this shows statistically significant differences at weeks 2 and 8 *(arrowed)*, with major trends towards similar significant differences in weeks 1 and 3 (Lacey and Crisp 1980)

Can These Drugs Be Helpful in the Treatment of Anorexia Nervosa?

The author's experience is that phenothiazine can sometimes be helpful during weight gain occurring during refeeding treatment in anorexia nervosa. This is especially so in those rare instances when the disorder is truly close to schizophrenia (Hsu et al. 1981). Antidepressant drugs are rarely of value in anorexia nervosa. Whilst their main and immediate effect seems to be to increase drive behavior (Crisp 1980a), at their worst they specifically promote and release carbohydrate craving. This is often seriously counterproductive in respect of treatment, since the anorectic becomes terrified of her loss of control, potential or actual, and either withdraws from treatment with her phobic avoidance stance reinforced or is likely to vomit defensively in relation to the increased food intake which is now otherwise beyond her control. Such developments, of course, may not arise within an initial controlled treatment program but they emerge thereafter, the anorectic having conceded the need to gain weight as the price of being allowed to escape from the treatment situation.

Depression and Anorexia Nervosa

A relationship between depression and anorexia nervosa has long been postulated. Many years ago Nemiah (1950) suggested that the condition was similar to agitated depression on the grounds of the patient's depressed mood, restlessness, and early morning wakening. Since then various workers, including

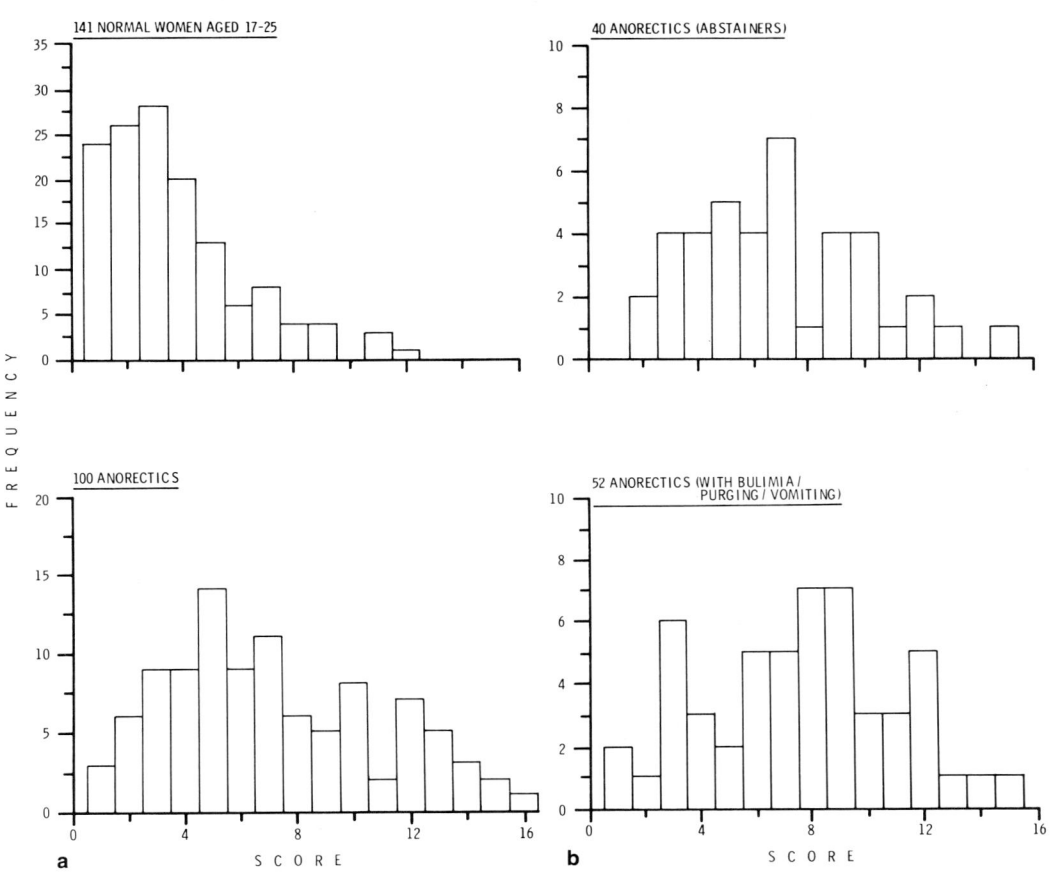

Fig. 4. a Scores on the depression scale (CCEI) in a population of 100 anorectics and 141 normals matched for age. The anorectics score significantly higher than the normals. **b** Comparison of 40 anorectic abstainers with 52 anorectic bulimics. The latter scored significantly higher but again there was great variability of scores, which were rarely in the range for severely depressed people (Crisp and Bhat 1982)

the present author and his colleagues (Kalucy et al. 1977; Crisp and Toms 1972), have found evidence of depressive disorder in excess in first-degree relatives. Some have drawn the conclusion that this renders it more likely that anorexia nervosa is, in fact, a depressive illness or a variant of it. This view has been proclaimed most recently in respect of bulimia nervosa (Hudson et al. 1982). I have always recognized the presence of depression within the anorectic syndrome. In mood terms, the anorectic is often sad and sometimes sporadically overwhelmed with guilt, disgust, and loathing for her body. She sometimes mutilates it and sometimes poisons herself with drug overdosage. With chronicity and with bulimia and defensive vomiting and purging and the sense of a more precarious adjustment that this brings, (Fig. 4), anorectics become increasingly depressed (Crisp and Bhat 1982). I also believe that this is a product

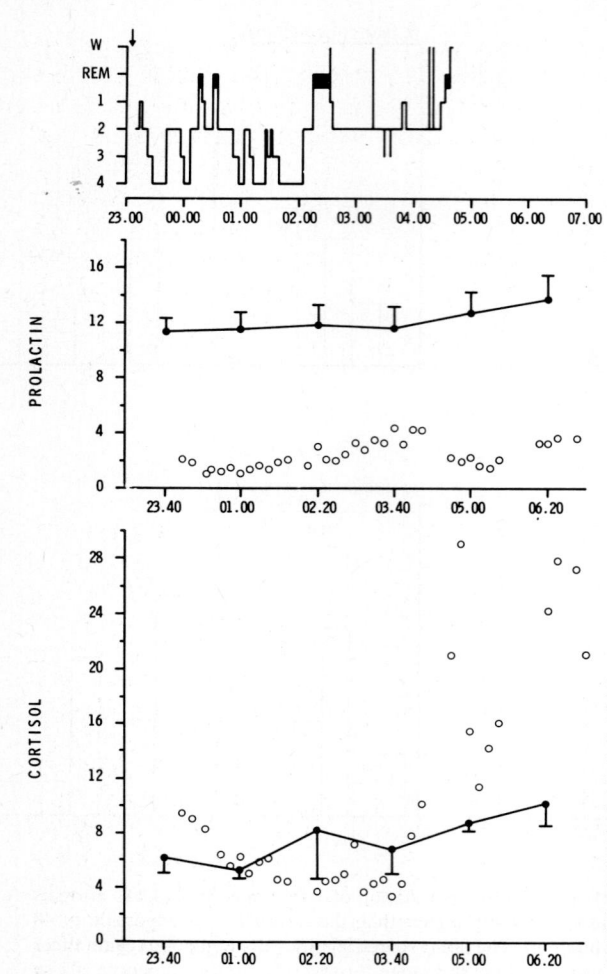

Fig. 5. Some typical sleep characteristics of a patient with anorexia nervosa, an 18-year-old female weighing 32.8 kg, height 1.58 m. Light, restless sleep with early final waking is evident in the EEG. This feature coincides with a massive increase in plasma cortisol levels to way above peak normal levels *(solid black line),* and a related and significant reduction in plasma prolactin levels (Crisp 1980a)

of their exhaustion and progressive isolation. Thus the depression in anorexia nervosa is similar to that found as a consequence within other syndromes involving maladaptive coping styles, e.g., alcoholism and agoraphobia. Moreover, the degree of depression in anorexia nervosa, as measured by a standardized rating scale, is about the same as that found in populations with anxiety state, anxiety phobic state, and schizophrenic and personality disorders, and is well below that found in depressive illness (Crisp et al. 1978; Ben-Tovim et al. 1979). This is all very different from suggesting that there is a primary disorder of depression within anorexia nervosa. However, most people who champion the idea of a primary depressive disorder point to the early morning wakening and dexamethazone nonsuppression as true indicators of underlying depression. I have identified early morning wakening as a characteristic of anorexia nervosa (Crisp 1967), often minimized by the person affected, in contrast to the state of affairs within depressive illness when the patient complains bitterly of early morning wakening, though often this is not very striking in terms of actual time of wakening (Crisp and Stonehill 1973; Crisp 1980a). The early morning wakening in anorexia nervosa is entirely related to the starvation and quite unrelated to mood (Crisp and Stonehill 1976). Furthermore, the failure of dexamethazone to suppress cortisol secretion is likely to be a function of the high arousal associated with starvation. Figure 5 shows the characteristic early morning wakening in a severely emaciated anorectic associated with a great surge of cortisol arising just before wakening. The impulse to waken and forage, experienced in a highly aroused state, rather than to continue to sleep unproductively, is not rooted in depression.

Psychopathology of Anorexia Nervosa

To consider psychopharmacological approaches to anorexia nervosa further we surely need to look beyond the above categories of drugs and the concept of depression to the complexities of the psychopathology of the condition.

Theories concerning the experiential basis of anorexia nervosa have to do with its being a reaction to adolescent maturational challenges, usually arising within a family context. To me at least, it appears to be a coping style arrived at unconsciously but defensively in relation to such problems. The emphasis is usually placed on the importance of the condition for sustaining a sense of competence and worth in the face of potential loss of control and devaluation. The morbid relationship of food to bodily experience and to feelings, first established in childhood, is intensified by growth and puberty. Over and above this I have emphasized the view that puberty is pivotal to the condition, which reflects a regression in relation to it and the consequent adoption of a phobic avoidance stance in relation to mature body weight. The anorectic's determination and obstinacy is not primarily compulsive in quality but defensive, and is born of her desperate search for relief from anxiety. She is like the hysteric and, for instance, the agoraphobic or the alcoholic, hemmed in and with no further defensive position to fall back on other than suicide. Under these circumstances she needs to dominate and control her circumstances and she is experienced as tyrannical and manipulative by those around her. In addition, anorectic families are characterized by conflict avoidance patterns, and in her anorexia nervosa the person with it protects her family from conflict. With puberty reversed, the stresses of sexual maturation are eliminated both for herself and for her family. Her new problem is to maintain her stance, biologically unstable though it is, because of the body's thrust for regrowth. Her family's problem now is to live with a self-absorbed uncommunicative anorectic − a visible indictment of their "failure", visible to all though obscure in its origins to them (Crisp 1980b).

Figures 6 and 7 outline some of these ideas and invite consideration of the role that drugs might play in restoring normality in the system. Figure 6 identifies four areas of evolution of the condition, and Fig. 7 further elaborates some of

	A	B		C	D
	Childhood (family)	Adolescence	(Outside World)	Regression	

A — Childhood (family)
Family attitudes to 'self', 'part self' sexuality, 'good' 'bad', conflict avoidance patterns reinforced/activated by puberty

Growth
Food
Puberty

B — Adolescence (Outside World)
Conflict
Increasing sensitivity about 'fatness' 'bad'
curbing fatness
dieting

Search for controls/identity

Female 'fatness'
emerging sexuality;
exploration of potential for adult peer relationships

C — Regression
Crisis
Panic
xs calorie restriction

Resolution of crisis

1) *Terror* of weight gain
2) *Starvation*, high arousal, impulse to gorge, hyperactivity, early morning waking, hoarding, preoccupation with food etc.

D
Sustained phobic avoidance posture

Anorexia nervosa

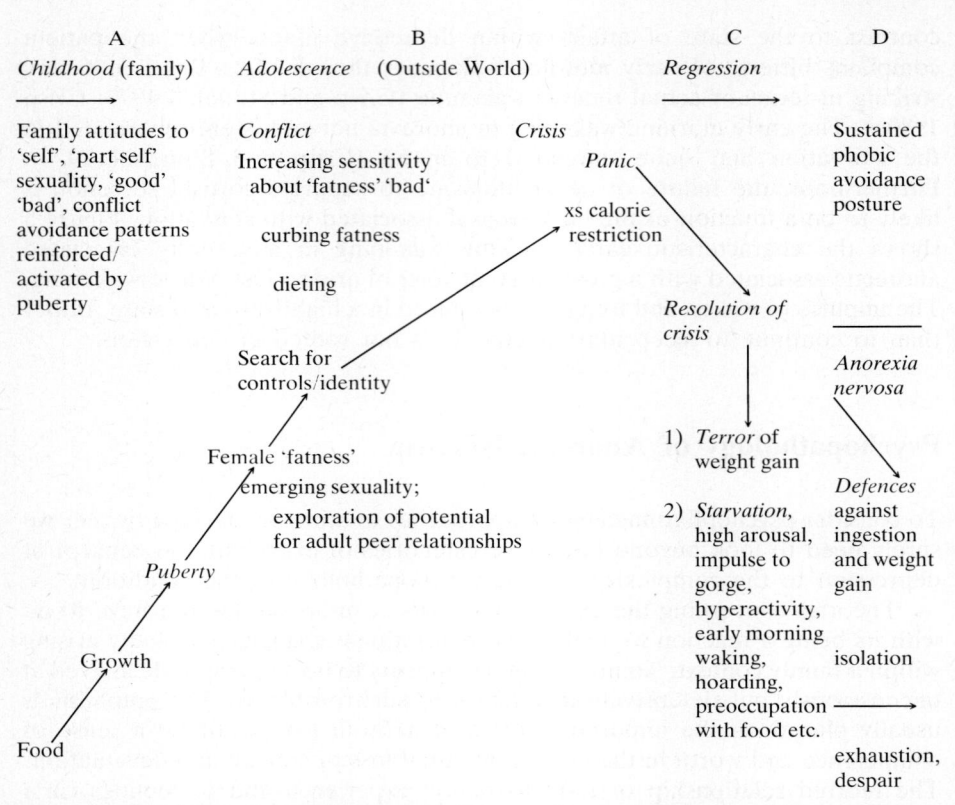

Defences against ingestion and weight gain

isolation

exhaustion, despair

Fig. 6. Aspects of the social and biological development of anorexia nervosa in childhood *(A)* early adolescence *(B)*, and the illness *(C* and *D)*

---

55 kg

Avoidance of normal body wt. ↓
fear of wt. gain

State of starvation (arousal, preoccupation with foraging and ingestion) ↑

30 kg

	Defences	
Feelings	*Against others*	*Self*
Panic	Denial	Ritual
Anger	Manipulation	Social avoidance
Disgust	Social avoidance	Vomiting
Ecstacy	Domination	Purging
Hopelessness	Apparent self sufficiency	Diuresis
Sadness		Vegetarianism
Low self esteem		Activity

Progressive isolation, exhaustion, despair

→

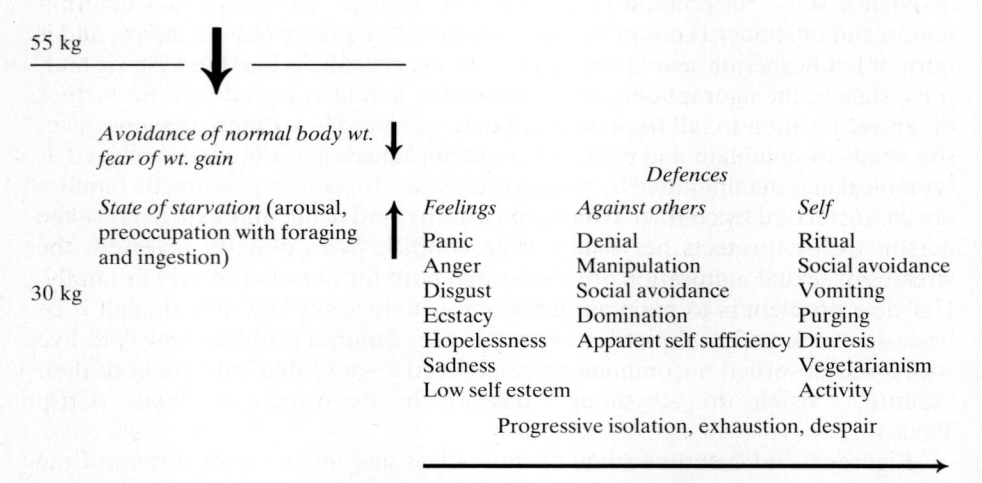

Fig. 7. Some of the feelings and defenses generated within anorexia nervosa against a background of the food-deprived body's impulse to ingest in the face of the anorectic's terror of weight gain

the defensive stances and distressing experiences in anorexia nervosa. Pharmacological strategies within this latter context are likely to be directed more at symptomatic relief and to some extent have been referred to earlier. However, I would like to focus on a strand of psychopathology that is identifiable within Fig. 6 and is more fundamental. Puberty is the bridge between columns A and B. It is forced upon us by Nature, given proper nourishment, freedom from infection, and compliance with nurturing. Ironically, it is indeed often generated precipitously by the overprotective mother. It is sometimes experienced as alien and indeed may be rejected almost immediately by the person who develops early-onset anorexia nervosa. Others tolerate it, dissipating their new and frightening energies in displacement activities punctuated by catastrophic forays into peer relationships, until they fall victim a few years later. The threat of puberty and the adolescent growth spurt within which we normally eat more than at any other time in our lives, is terrifyingly reactivated for many anorectics (column C), who are doomed to live on the brink of or even within this state within their illness. The threat within them to grow and hence to eat leaves them with an ever-increasing sense of loss of control, initially of weight but essentially of almost undifferentiated impulses. Perhaps a drug that curbs appetite under these circumstances would give the anorectic a sense of greater control and paradoxically permit her to eat more (Crisp 1978). In the past amphetamine and similar compounds have been obvious candidates for such a function, and indeed we started a trial of such medication a few years ago but had to abandon it. We had felt that those with the bingeing/vomiting syndrome would be the most likely to be helped, but it did not seem that they were dramatically improved, and their vomiting meant that the absorption of the drug was somewhat variable! More recently Trenchard and Silverstone (1983) have shown that naloxone, an opiate antagonist, reduces food intake in normal subjects. Perhaps, on the grounds advanced above, this is why Moore et al. (1981) found that it helped their anorectics to gain weight, or perhaps this was due to some other aspect of the treatment setting, or to reduced vomiting under such treatment. Silverstone (1983) has recently reviewed the pharmacology of zimelidine, another antidepressant, and also amitriptyline in relation to their clinical effects. The former leads to weight loss and the latter, as previously implied, to weight gain in depressed subjects. Silverstone points out that the former is believed to act by blocking the reuptake of 5HT into presynaptic neurons, whilst amitryptyline has been reported to have a pronounced blocking action of postsynaptic 5HT receptors which, he says, could explain appetite-stimulating properties in depressed subjects. Perhaps once again, the report of zimelidine enabling anorectics to gain some weight has its origins in their paradoxical response rooted in their abnormal fears of weight and food. I believe that such drugs may have a limited role in the treatment of anorexia nervosa if the person affected is not effectively engaged in a therapeutic relationship. They may then enable her to sustain her present low weight rather than to go on losing weight as an insurance policy. They may help her gain some weight over a limited period when this is the price she has to pay to escape from the treatment situation. But I do not believe that they improve the longer-term prognosis.

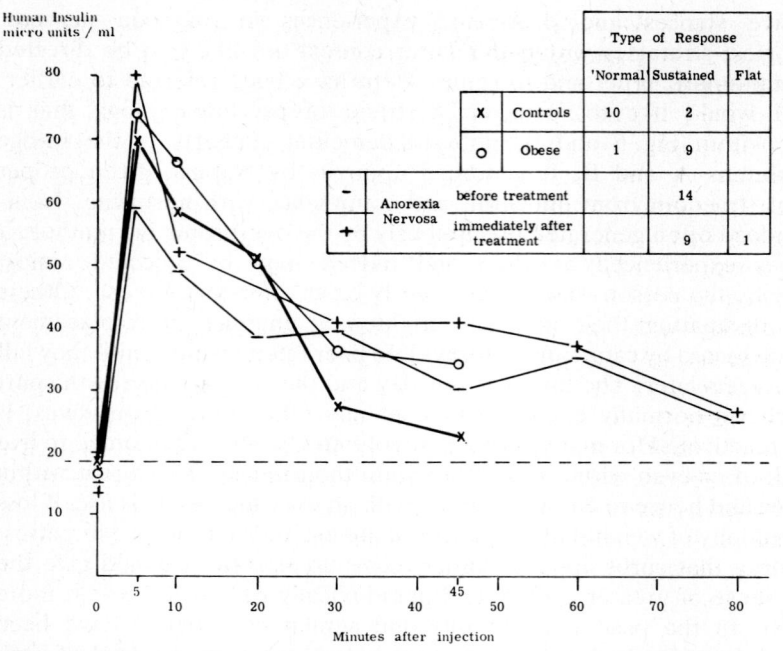

Fig. 8. Insulin response to a rapid IV injection of 25 g glucose (Crisp 1967)

Let us now look at an aspect of the chemistry of anorexia nervosa — the chemistry of the fat cell and the insulin response to a glucose load. Anorectics show a sustained insulin response to an IV glucose load (Crisp 1967) (Fig. 8). Resting insulin levels are high in anorectics who have only just begun to refeed (Kalucy et al. 1976). (Insulin, of course, is quite the wrong drug to give anorectics. The idea in the past has been that it induces hypoglycemia and thereby renders the anorectic hungry. In fact, of course, it simply renders her even more hungry and fearful, and once again short-term bingeing and apparently gratifying weight gain may be replaced by long-term defensive vomiting and avoidance of further treatment.) In the past I have suggested that these potentially high levels and the sustained response are a basis for the potential addiction to carbohydrate and the perceptual disorder as well as the periodic amenorrhea that characterizes weight-conscious and dieting adolescent females including anorectics (Crisp 1967; Crisp and Kalucy 1974).

Does the anorectic's fear that if she eats any carbohydrate she will not be able to stop, in any way reflect this abnormal process that such ingestion generates within her? Does starving yourself of carbohydrate generate an addiction for it? Over the years my own treatment programs of anorexia nervosa and obesity have been rooted in this notion. Our anorectics are expected to eat moderate amounts of carbohydrate as also, when we were treating them by diet and psychotherapy, were our obese patients (Crisp and Stonehill 1973).

There is a relevant literature on this involving sleep, wherein sleep seems to provide the 24-h modulator or at least to be an indicator and predictor of carbohydrate intake. We have shown a relationship between carbohydrate intake and the amount of subsequent REM sleep the following night (Phillips et al. 1975). Tryptophan as a 5HT precursor may be the mediator. For instance we have also found that REM sleep is related to plasma free tryptophan levels. Meanwhile Wurtman and Fernstrom (1975) have demonstrated such links within the mammal. They found that the administration of tryptophan, injection of insulin, or consumption of protein-free carbohydrate increased brain tryptophan, serotonin, and its metabolites. They concluded that food can influence cerebral metabolism, and Wurtman (1983) has recently elaborated on this. Meanwhile there is some evidence that the amount of REM sleep bears an inverse relationship to the next day's carbohydrate intake (Siegel 1975), i.e., in "normal" mammals high carbohydrate intake one day leads to low carbohydrate intake the next day. Such self-regulation round a set point is known to exist. In REM sleep we seem to be tapping the central mechanism, and in tryptophan, one of the chemical mediators. So far as anorexia nervosa is concerned I suggest that moderate carbohydrate intake is the way to curb the otherwise endless impulse to consume it excessively. The "drug" treatment for anorexia nervosa is moderate and controlled carbohydrate ingestion fascilitated by a therapeutic relationship that allows the anorectic to foresee the possibility of an alternative coping style in life.

Summary

Anorexia nervosa is viewed as having its roots in adolescent existential problems. The person affected does not see him- or herself as afflicted despite the fact that the condition may be life-threatening. Patients may well not comply with prescribed treatments. Consumption of certain psychotropic drugs can be helpful to survival, but others are harmful and can lead to the decompensation of a relatively stable anorectic stance. No drugs have been demonstrated as effective in promoting permanent recovery and autonomy; nor is it likely that such a drug will emerge. However, if anorectics can be persuaded that their craving and fear of loss of control over ingestion is partly a product of this extreme avoidance of carbohydrate, then their subsequent moderate consumption of it, under controlled treatment conditions, sometimes dilutes their fear in this respect and permits weight gain. Meanwhile the best that can be hoped for in the short term is that they construe carbohydrate and other highly calorific food as a drug and not a poison.

References

Amdisen A (1964) Drug-induced obesity: experiences with chlorpromazine, perphenasine and clopenthixol. Dan Med Bull 11: 182–189

Ben-Tovim DI, Marilov V, Crisp, AH (1979) Personality and mental state (PSE) within anorexia nervosa. J Psychosom Res 23: 321–325

Crisp AH (1965) A treatment regime for anorexia nervosa. Br J Psychiatry 112: 505—512
Crisp AH (1967) The possible significance of some behavioural correlates of weight and carbohydrate intake. J Psychosom Res 11: 117—131
Crisp AH (1978) Disturbances of neurotransmitter metabolism in anorexia nervosa. Proc Nutr Soc 37: 201—209
Crisp AH (1980a) Sleep, activity, nutrition and mood. Br J Psychiatry 137: 1—7
Crisp AH (1980b) Anorexia nervosa: let me be. Academic Press, London; Grune and Stratton, New York
Crisp AH, Bhat BV (1982) 'Personality' and anorexia nervosa — the phobic avoidance stance. In: Krakowski AJ, Kimball CP (eds) Psychosomatic medicine in a changing world. Karger, Basel, pp 178—200
Crisp AH, Roberts FJ (1962) A case of anorexia nervosa in a male. Postgrad Med J 38: 350—353
Crisp AH, Stonehill E (1973) Aspects of the relationship between sleep and nutrition: a study of 375 psychiatric out-patients. Br J Psychiatry 122: 379—394
Crisp AH, Stonehill E (1976) Sleep, nutrition and mood. Wiley, London
Crisp AH, Toms DA (1972) Primary anorexia nervosa or weight phobia in the male. Report on 13 cases. Br Med J 1: 334—338
Crisp AH, Jones GM, Slater P (1978) The Middlesex Hospital Questionnaire: a validity study. Br J Med Psychol 51: 269—280
Dally P, Sargant W (1960) A new treatment for anorexia nervosa. Br Med J 1: 1770—1773
Halmi KA, Eckert E, Falk JR (1983) Cyproheptadine, an antidepressant and weight-inducing drug for anorexia nervosa. Psychopharm Bull 19: 103—105
Hudson JI, Laffer PS, Harrison GP (1982) Bulimia related to affective disorder by family history and response to dexamethasone suppression test. Am J Psychiatry 139: 685—687
Hsu LKG, Meltzer E, Crisp AH (1981) Schizophrenia and anorexia nervosa. J Nerv Ment Dis 169: 273—276
Kalucy RS, Crisp AH, Chard T, McNeilly A, Chen CN, Lacey JH (1976) Nocturnal hormonal profiles in massive obesity, anorexia nervosa and normal females. J Psychosom Res 20: 595—604
Kalucy RS, Crisp AH, Harding B (1977) A study of 56 families with anorexia nervosa. Br J Med Psychol 50: 381—395
Lacey JH, Crisp AH (1980) Hunger, food intake and weight: the impact of clomipramine on a refeeding anorexia nervosa population. Postgrad Med J [Suppl 1] 56: 79—85
Moore R, Mills IH, Forster A (1981) Naloxone in the treatment of anorexia nervosa: effect on weight gain and lipolysis. J R Soc Med 74: 129—131
Nemiah JC (1950) Anorexia nervosa — a clinical psychiatric study. Medicine 29: 225—268
Paykel ES, Mueller PS, de la Vergne PM (1973) Amitriptyline, weight gain and carbohydrate craving: a side effect. Br J Psychiatry 123: 501—507
Phillips F, Chen CN, Crisp AH, Koval J, McGuiness B, Kalucy RS, Kalucy EC, Lacey JH (1975) Isocaloric diet changes and EEG sleep. Lancet 2: 723—725
Siegel JM (1975) REM sleep predicts subsequent food intake. Physiol Behav 15: 399—403
Silverstone T (1983) The clinical pharmacology of appetite — its relevance to psychiatry. Psychol Med 13: 251—253
Silverstone T, Schuyler D (1975) The effect of cyproheptadine on hunger, calorie intake and body weight gain. Psychopharmacology 40: 335—340
Singh MM, Vergel De Dios L, Kline NS (1970) Weight as a correlate of clinical response to psychotropic drugs. Psychosomatics 11: 562—570
Trenchard E, Silverstone T (1983) Naloxone reduces food intake of normal human volunteers. Appetite J Intake Res 4: 43—50
Vendsborg PB, Bech P, Rafaelson OJ (1976) Lithium treatment and weight gain. Acta Psychiatr Scand 53: 139—147
Wurtman RJ (1983) Behavioural effects of nutrients. Lancet 1: 1145—1147
Wurtman RJ, Fernstrom JD (1975) Control of brain monoamine synthesis by diet and plasma amino acids. Am J Clin Nutr 28: 638—647

The Basis of Naloxone Treatment in Anorexia Nervosa and the Metabolic Responses to It

I. H. Mills and L. Medlicott[1]

Clinical Assessments

Some patients with anorexia nervosa are somewhat overweight prior to their illness and they give a history of initial dieting to reduce their weight. In a number of cases their food reduction is initially not severe and they have gradual weight loss without any interference with menstruation. After a period of perhaps a few months, they describe a relatively sudden change in their attitude to food; frequently they use the same words to describe this change: "Suddenly the diet took over." They mean by this that they no longer feel they are in command of their eating habits and they feel as if something inside them is determining more dramatic reduction in food intake and consequently in weight also. From this point on, they frequently have no more menstrual periods.

Our studies some 10 years ago (Mills et al. 1973) investigated precisely what such girls were doing when their severe dieting first started. The results showed that in 75% of cases they were working for important examinations. Among the remaining 25% many were tackling difficult social problems, such as trying to prevent their parents separating or, in one case, trying to teach her siblings with dyslexia to read and write. Whether the main challenge was an academic one or a social one, the response of the girl who is liable to become anorectic is different from that of most other girls. She feels that once she has embarked upon tackling a problem she must not give up until she has succeeded. She is not satisfied with her achievements and drives herself on to reach higher goals. When working for examinations she concentrates on what she does not know, without asking herself if she knows what is expected of her. In this sense she is a perfectionist and once she becomes caught up in anorexia nervosa she becomes even more of a perfectionist, at least in the aspects of her life that she is pursuing so intently.

The number of patients being seen with anorexia nervosa has undoubtedly increased in the last 20 years. This stems from the time in the early 1960s when the slim image for the human female became widely accepted and was epitomized in the film of Twiggy. As a result of this slim image becoming dominant, from time to time a whole group of girls or young women will decide to go on a diet together. By the end of about a week most of them give up dieting because they cannot stand being hungry. However, the one or two who do not give up are self-selected for these qualities of perfectionism and determination to strive for the goal they had set themselves.

---

1 University of Cambridge, Department of Medicine, Addenbrooke's Hospital, Hills Road, Cambridge CB2 2QQ, Great Britain

The Psychobiology of Anorexia Nervosa
Edited by K. M. Pirke and D. Ploog
© Springer-Verlag Berlin Heidelberg 1984

Not all young women with a potentiality to become perfectionists would respond to this dieting challenge. They are more likely to do so when the circumstances of their lives are such as to get their mental arousal into a high state. Inability to solve their family and social problems becomes a continuous challenge to them; or when working for examinations (which may still be 6 or 9 months away) they may be driven by the aspirations for them of their parents, their teachers, their peer group, and themselves. Their obsessional, perfectionist nature causes them to respond excessively to the challenges with the result that they work with greater intensity or for longer hours, or both, and thereby raise their mental arousal level. In the Yerkes-Dodson curve relating mental arousal to efficiency, higher arousal is rewarded by higher achievement until the top of the curve is reached. Beyond this point, increased arousal leads to a downward path of achievement. Only those with great determination can drive arousal beyond the top of the efficiency curve: most people become fatigued and give up before this point is reached. The point at which the top of the curve is reached depends upon the task undertaken (Poulton 1970) and probably the personality of the individual.

The potential anorectic young woman is more likely to respond excessively if the dieting challenge happens to come at the time when she is in a high arousal state. Having decided to go on a diet with the group, the self-selected, determined girl will not give up just because of the hunger. The more she persists the more likely she is to get through this hunger barrier and, in her obsessional state, she is more likely to increase the severity of the dietary restrictions.

By whichever route the girl gets caught up in dieting, she reaches the point at which the diet takes over. At this point she becomes driven by some internal force which has the quality of a compulsive mechanism. The more she dwells on the dieting, the more the thoughts of food and weight and size and diet and size of clothes become a continuous stream in her brain. Frequently such girls think of food they would enjoy eating and actually prepare it for the family but the compulsive drive in the brain prevents them from eating the foods they prepare. Though they think of food, they have lost their normal hunger control mechanism.

The thoughts of food, weight, diet, and size become relentless in their repetition. The girls with insight appreciate that these continuous repetitive thoughts are like a "gramophone stuck in one groove" playing a repetitive part of a tune. They may even plead for some way of escaping from this never-ending series of repetitive thoughts. Their own device for coping with it is to concentrate on some interesting and demanding task. This shuts out the gramophone but only for so long as they maintain their concentration. This potentiates their exhaustion and may make working become a compulsion in its own right.

The compulsive nature of the drive in their brains becomes obvious in their work which they are unwilling to give up. It becomes even more obvious in the 20%−30% who swing over to compulsive eating. To observe them or listen to their description of the relentless urge to binge is to understand how intense the internal drive is. When their food supply is restricted they will go almost to the same lengths to get food that a heroin addict will go to get heroin. They know the bingeing is wrong; they feel intensely worried about the effect on their weight;

they feel guilty at what they are doing and disgusted with themselves when they vomit: yet they are helpless to resist the compulsive drive. They cannot understand why their intelligence cannot stop their bingeing. They resemble the pure anorectic in being unable to differentiate between the thoughts related to the compulsive, starving process and their intelligent, rational thinking. In desperation, the bingeing, anorectic girl is glad if overeating leads to vomiting and rapidly trains herself to vomit after eating. Eventually this also becomes a compulsion and vomiting after quite normal meals then occurs.

It is these facts, told over and over again as each new patient presents, that have convinced us of the compulsive nature of anorexia nervosa. The reward from starvation when tackling severe challenges that they feel they must meet is to be found in the studies of starvation by professional fasters at the beginning of the century (Benedict 1915). The effect on sleep was well recounted then and fits in with the sleep disturbance in anorectic patients. Benedict also studied mental ability and found that it improved during fasting to reach a plateau about the 14th day. The brain was faster and more accurate in solving problems. Though the anorectic girl may not consciously realize this, she is very likely to find the reverse effect if family and friends persuade her to eat more normally. This tends to lower arousal level, even to reveal a masked depression in some cases, but most of all it makes the girl's self-set goals much harder to reach. At this point she then returns to severe dieting with renewed determination, reinforced by her experience. Once she develops the distorted body image in her brain, she works even harder to "starve" it down to size.

Drugs in Relation to the Compulsive Drive in Anorexia Nervosa

If the mechanism in anorexia nervosa is a compulsive one, associated with the patient's self-motivated excessive drive to achieve her perfectionist goals, the questions which then arise are: (a) Where in the brain is the compulsive mechanism situated? (b) What neuronal pathways are involved? (c) Can we influence those pathways to the benefit of the patients? and (d) Can we demonstrate the compulsive drive in the patients?

With regard to the site in the brain where compulsive behavior might be located, one possibility is the hippocampus. It is through this bilateral structure that all new information must pass to be stored in intermediate and long-term memory (Popper and Eccles 1977). This would make it appropriate in individuals coping with academic or social challenges. The relentless repetition of thoughts of food, weight, diet, size, and related things could be related to the unique feature of the hippocampus known as "long-term potentiation". When stimulated beyond a certain point the hippocampus subsequently responds with a prolonged and exaggerated response. This potentiation has been shown to persist for at least 3 months in animals.

The hippocampus is stimulated by enkephalin (by disinhibition) and contains encephalinergic neurons which arise from the lateral entorhinal/perirhinal cortex. In addition, many hippocampal neurons are stimulated by the opioid dynorphin, which has been demonstrated in various cells in the hippocampus. Dynorphin stimulation produces long-lasting excitation (Henriksen et al. 1981) and may be the basis of the hippocampal long-term potentiation.

In a description of the coping mechanism in the brain (Mills 1981) we have indicated the important role played by the noradrenergic pathways in the brain. This being so, one might expect that drugs which block noradrenaline reuptake might be potentially valuable in facilitating the coping mechanism in patients. Certainly the young women who become motivated to fight against the compulsion to starve themselves find that the fight makes them exhausted and depressed. They then give up the attempt but when treated with nortriptyline or amitriptyline will put up a more prolonged effort to refeed themselves.

Opiates and Noradrenaline

The recent work of Yamamoto et al. (1981) on the effect of kyotorphin (Tyr-Arg) in the brain may also have a bearing on this. Kyotorphin excites cells in the cerebral cortex but its action is not reversed by naloxone: the inhibitor of this mechanism is noradrenaline. Since it is known that parts of the cerebral cortex, when stimulated, increase arousal in the reticular formation from above (for refs. see Mills 1981), it is quite likely that kyotorphin is the mechanism responsible for this and might be the means by which excessive arousal drives efficiency over the top of the Yerkes–Dodson curve. If these suggestions were correct, noradrenaline potentiation by such drugs as nortriptyline might be effective in lowering arousal till it reaches the efficient part of the curve. The sedative effect of such antidepressants is compatible with such lowered arousal.

Phenoxybenzamine, an adrenergic antagonist, in doses which did not produce too great a fall in blood pressure, appeared to have no good or bad effect.

Anticholinergics

Both nortriptyline and amitriptyline have anticholinergic actions, and any benefit of the drugs might reflect their action on cholinergic pathways. We therefore tested the cholinergic antagonist biperiden and in a number of patients increased sedation was reported. It did not, however, block the compulsive drive.

Dopamine

The early successful use of cholorpromazine by Dally (1969) might suggest that blockage of dopaminergic pathways would be advantageous. In our experience this drug is not effective in the more depressed anorectics and we mostly abandoned it in favor of antidepressants (Mills 1976).

However, the report that amantadine (a dopamine agonist) could block stereotyped behavior (Schneiden and Cox 1976) in animals led us to test it in some patients. One who had both anorexia nervosa and obsessional-compulsive neurosis with numerous rituals initially responded dramatically with complete cessation of rituals. She no longer responded when she relapsed and no other patients showed a beneficial response.

Some patients with high prolactin levels have stated that the dopamine agonist bromocriptine appreciably reduced and sometimes controlled compulsive eating. In other patients no benefit was noted.

5-Hydroxytryptamine

The reported success with clomipramine in patients with obsessional-compulsive neurosis suggested that it might be effective in anorexia nervosa. In doses up to 300 mg/day we have seen no significant improvement, and certainly no decrease in the compulsive thinking about eating, diet, weight, size, etc.

Scoring the Compulsive Drive

A coping questionnaire has been devised, consisting of 76 questions. This questionnaire has been validated and scored, and it is possible to extract from it scores for compulsions, for alcohol dependence, and for depression. The patients have been divided into pure anorectics, anorectics who have phases of compulsive eating ("bingeing anorectics"), patients with other compulsions (especially compulsive working, compulsive smoking, etc.), patients with depression and controls. The control group were matched for age, sex, social class, and responses to the Eysenck Personality Questionnaire.

Table 1 gives the scores for the various groups. The pure anorectics and the bingeing anorectics have highly significant compulsion scores compared with the controls, whereas the depressed patients do not. The bingeing anorectics come

Table 1. Compulsion, alcohol, and depression scores (mean ± SE) (Medlicott and Mills, to be published)

Group	n	Compulsion score	Alcohol score	Depression score
Controls	42	13.1 ± 1.1	0.7 ± 0.3	7.9 ± 0.9
Pure anorectics	34	32.1 ± 1.9**	0.6 ± 0.3	13.8 ± 0.8**
Bingeing anorectics	28	35.8 ± 1.9**	5.8 ± 1.2**	17.0 ± 0.6**
Other compulsives	28	28.0 ± 2.2**	1.6 ± 0.6	11.1 ± 1.1*
Depressed	30	14.8 ± 1.4	2.2 ± 0.6*	17.8 ± 0.7**

* $P < 0.01$
** $P < 0.005$

out with a very highly significant attachment to alcohol, which is shown neither by the pure anorectics nor by those with other compulsions. Depressed patients had a less severe but still very significant dependence upon alcohol.

The bingeing anorectics appear to have the highest compulsion score, which fits in with what we know of them clinically, but the pure anorectics are close behind them and also have a score very markedly different from the that of the control group.

The bingeing anorectics were the only group to have a high score on compulsive vomiting (9.6 ± 0.7 vs control values of 0.2 ± 0.1). All the patient groups had highly significant scores for the questions related to difficulties in social relationships (6.8−9.4, compared with control values of 4.0 ± 0.4).

The depression score was very high in all groups except those with "other compulsions", but even they were significantly different from controls ($P < 0.01$).

Treatment with Naloxone

Having failed to control compulsive behavior with drugs affecting neuronal transmission by noradrenaline, acetylcholine, dopamine, and 5-hydroxytryptamine, we decided that the opiate mechanisms in the brain might be responsible for the compulsive behavior and tried to inhibit them by naloxone infusion.

Naloxone was given by continuous IV infusion: in the first series it was given for times up to 61 days at doses between 1.6 mg and 6.4 mg per day (Table 2). Heparin, 200 U per 12 h, was added to decrease thrombosis and the tendency to staphyococcal septicemia. All patients were receiving, 50−200 mg amitriptyline per day and anticonvulsants. In spite of this, three patients had epileptic fits before the dose of anticonvulsants was increased.

Preliminary studies in diabetes mellitus (Ghosh et al. 1977) had indicated that naloxone facilitated reduction of plasma NEFA and β-hydroxybutyrate. In each anorexic patient ingestion of 3,000 kcal per day had been urged for some

Table 2. Dosages of naloxone and duration of IV infusion

Dose (mg/day)	Duration (days)
1.6	19
3.2	20
4.8	20
4.8	31
4.8	35
6.4	35
6.4	38
4.8	58
4.8	61

time before the infusion of naloxone, and the same diet was continued during and after the infusions to assess the effect on weight gain. [These results have been previously reported by Moore et al. (1981).] Blood was taken during fasting several times a week to estimate NEFA and β-hydroxybutyrate in three patients. The responses are shown in Fig. 1. The second patient (Fig. 1B) initially received infusions of saline and did not know when she was changed to naloxone. There was no fall in NEFA or β-hydroxybutyrate during the saline infusion, but in each case there was progressive depression of fasting levels of NEFA and β-hydroxybutyrate. In the patient shown in Fig. 1C there was a short phase of increased levels of these substances in the plasma early on in the infusions; the reasons for this are unknown.

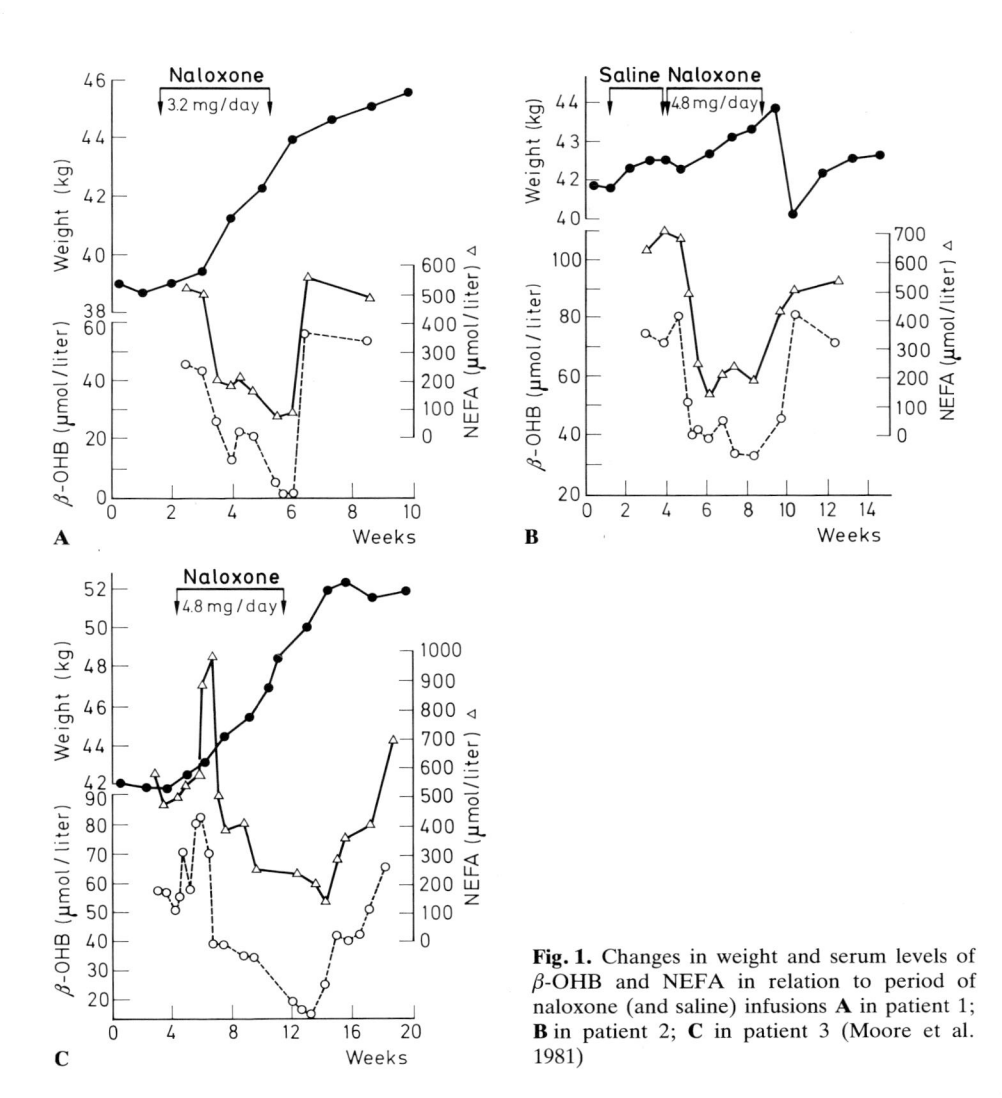

Fig. 1. Changes in weight and serum levels of β-OHB and NEFA in relation to period of naloxone (and saline) infusions **A** in patient 1; **B** in patient 2; **C** in patient 3 (Moore et al. 1981)

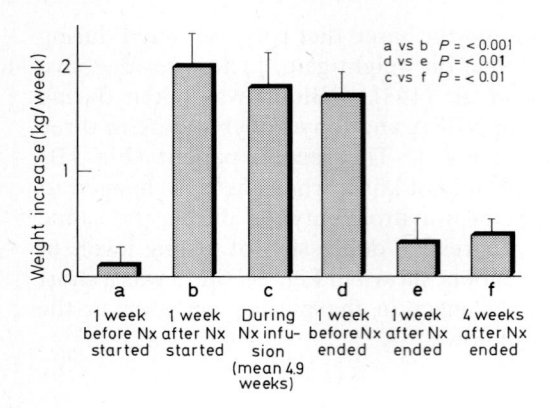

Fig. 2. Weight gain before, during, and after naloxone (Nx) infusion in 12 patients with anorexia nervosa (mean ± SE) (Moore et al. 1981)

Bar labels:
a: 1 week before Nx started
b: 1 week after Nx started
c: During Nx infusion (mean 4.9 weeks)
d: 1 week before Nx ended
e: 1 week after Nx ended
f: 4 weeks after Nx ended

a vs b $P = <0.001$
d vs e $P = <0.01$
c vs f $P = <0.01$

On cessation of naloxone the plasma NEFA and β-hydroxybutyrate levels rapidly rose to pretreatment levels. The mean responses in terms of weight of 12 patients treated with doses of naloxone up to 6.4 mg per day are shown in Fig. 2. The marked rise in weight gain per week began in the first week of treatment and was maintained throughout the treatment period. Weight gain decreased markedly on cessation of naloxone infusion.

From the marked increase in weight gain with no change in food intake, together with the depression of plasma levels of NEFA and β-hydroxybutyrate, it seems most probable that the naloxone effect was due to inhibition of lipolysis. One must assume that opiate-determined lipolysis is one of the factors acting against weight gain during the ingestion of 3,000 kcal per day prior to the naloxone infusion (Moore et al. 1981).

Changes in mental attitude to weight gain were only gradual during naloxone infusion, and reduction of the dose frequently led to resurgence of anorectic ideas. In later studies naloxone was always reduced gradually and as a result several patients were left with evidence of compulsive behavior for a while each time the naloxone dosage was reduced. The most effective responses, with the lowest relapse rate, were in those who received infusions continuously for 3 months or more.

The results of naloxone infusions, recorded 6 months after discharge from hospital, with the doses and duration of infusions are shown in Fig. 3. In two patients the clinical state was quite unsatisfactory at the times they decided to discharge themselves, despite long periods of naloxone infusion. One patient had only a short infusion and quickly relapsed. One patient had only barely stopped compulsive eating, though she had completely stopped compulsive vomiting. Soon after she was discharged at her own request, and after rapid withdrawal of naloxone, she had a resurgence of intense compulsive eating but could not make herself vomit. Her weight gain was consequently alarming.

Eight of the patients were in an excellent and stable state 6 months after discharge.

Although the initiation of the compulsive behavior in anorexia nervosa may be driven by the action of opiates in the brain, the continued performance of the

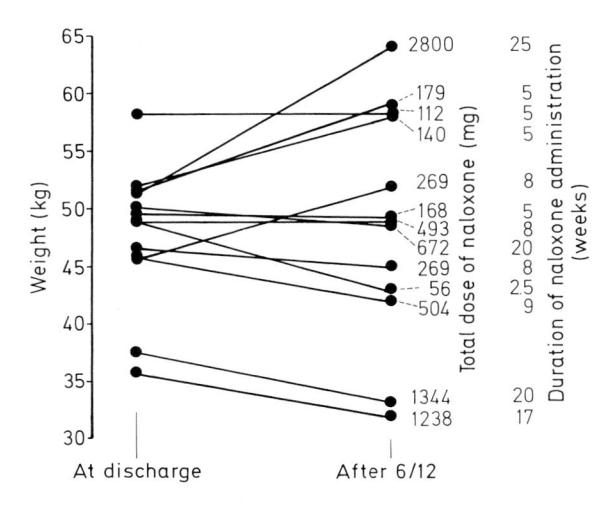

Fig. 3. The weights of 13 anorectic patients at the time of discharge from hospital and 6 months later, together with the total doses of naloxone they had received and the duration of naloxone administration

behavior must become imprinted in the memory mechanism. Under some circumstances opiates have been shown to enhance memory consolidation (Kovacs and de Wied 1981). In studies in rats the opiate kyotorphin inhibited extinction of avoidance response while, conversely, naloxone facilitated extinction of the response (Yamamoto et al. 1982). Naloxone does not appear to extinguish the memory in the patients treated by infusions, except while the infusion is continuing. If extinction of memory is part of its action, it appears that resurgence of memory can also occur. This might explain the well-known relapses of anorectics, which are particularly likely to occur when they are under the stress of an open-ended commitment. It is under these circumstances that their potential to drive themselves excessively is most likely to lead to a new phase of anorexia. Whether this is a conditioned response, the renewed initiation of the same behavior as previously, or the reappearance of a memory which appeared to have been extinguished is difficult to tell.

Summary

Studies in patients with anorexia nervosa have shown that they have frequently embarked upon open-ended commitments, especially academic and social ones. Increased frequency of anorexia nervosa is assumed to be related to the acceptance of the slim image for females over the last 20 years.

Girls embarking on dieting mostly give up when too hungry: potential anorectic patients are self-selected for perfectionist attitudes and determination to pursue goals they have set. They may end up on the downward part of the Yerkes-Dodson curve relating arousal to efficiency. Their likelihood of so doing is related to their arousal state when they start dieting, which is dependent upon driving themselves in open-ended commitments.

Dieting suddenly becomes compulsive at a critical point and then such patients cannot resist it. Menstruation then usually stops.

Compulsive thinking about diet and weight-related subjects tends to persist in the brain when it is not intensely occupied with other matters.

Severe dieting produces the same arousal and increased brain efficiency as was described early in the century in professional fasters. Return to eating may uncover masked depression and impair efficiency so that patients return to dieting.

The compulsive nature of anorexia nervosa led to a search for drugs to interfere with neural pathways which might be involved. No success was found with traditional neuronal transmitters so opiate stimulation was considered. The site of such action and the basis of compulsive behavior could be the hippocampus.

A coping questionnaire was developed and validated. It showed high compulsion scores in pure anorectics and those who also had phases of compulsive eating. The latter group also, uniquely, had high alcohol dependence scores.

Naloxone, the opiate antagonist, when administered by continuous IV infusion, led to depression of plasma nonesterified fatty acids and β-hydroxybutyrate, associated with more rapid weight gain with the same diet.

Prolonged infusions were necessary to abolish the compulsive mechanisms. Gradual withdrawl was necessary to prevent relapse. Naloxone might, in part, act by helping to extinguish the learned behavior of using starvation to raise arousal to cope with open-ended commitments.

References

Benedict FC (1915) A study of prolonged fasting. Carnegie Institution of Washington, Washington
Dally P (1969) Anorexia nervosa. Heinemann, London
Ghosh P, Mills IH, Moore R (1977) The anti-ketotic effect of opiate receptor blockade by naloxone infusion in human diabetes. J Physiol 271: 51–52P
Henriksen SJ, Chouvet G, McGinty J, Bloom FE (1982) Opioid peptides in the hippocampus: anatomical and physiological considerations. In: Verebey K (ed) Opioids in mental illness. Ann NY Acad Sci 398: 207–219
Kovacs GL, de Wied D (1981) Endorphin influence on learning and memory. In: Martinez JL, Jensen RA, Messing RB, Rigter H, McGaugh JL (eds) Endogenous peptides and learning and memory processes. Academic Press, New York, pp 231–247
Mills IH (1976) Amitriptyline therapy in anorexia nervosa. Lancet 25: 687
Mills IH (1981) The coping mechanism. In: Edholm OG, Weiner JS (eds) Principles and practice of human physiology. Academic Press, London, pp 425–449
Mills IH, Wilson RJ, Eden MAM, Lines JG (1973) Endocrine and social factors in self-starvation amenorrhoea. In: Robertson RF (ed) Symposium – anorexia nervosa and obesity. Royal College of Physicians of Edinburgh, Edinburgh, pp 31–43 (publication number 42)
Moore R, Mills IH, Forster A (1981) Naloxone in the treatment of anorexia nervosa: effect on weight gain and lipolysis. J R Soc Med 74: 129–131
Popper KR, Eccles JC (1977) The self and its brain. Springer, Berlin Heidelberg New York
Poulton EC (1970) Environment and human efficiency. Thomas, Springfield

Schneiden H, Cox B (1976) A comparison between amantadine and bromocriptine using the stereotyped behaviour response test (SBR) in the rat. Eur J Pharmacol 39: 133–141

Yamamoto M, Kawamuki K, Satoh M, Tagaki H (1981) Excitatory action of microelectrophoretically applied kyotorphin (Tyr-Arg) on unitary activity in the rat cerebral cortex. In: Takagi H, Simon EJ (eds) Advances in endogenous and exogenous opioids. Kodansha, Tokyo, pp 220–222

Yamamoto M, Kawamuki K, Satoh M, Takagi H (1982) Kyotorphin (1-Tyr-1-Arg) inhibits extinction of pole-jumping avoidance response in the rat. Neurosci Lett 31: 175–179

PET Investigation in Anorexia Nervosa: Normal Glucose Metabolism During Pseudoatrophy of the Brain

H. M. Emrich, J. J. Pahl, K. Herholz, G. Pawlik,
K. M. Pirke, M. Gerlinghoff, W. Wienhard, and W. D. Heiss[1]

Introduction

The recently developed technique of positron emission tomography (PET), in which fluorodeoxy-glucose is used as a label for the activity of glucose metabolism (Reivich et al. 1979), offers a unique possibility of investigating regional glucose metabolism in the living brain of humans (Brodie et al. 1983). Up to now, the method has been applied in neuropsychology (Phelps 1981), in different types of neurological disorders (Barrio 1983), and in major psychoses [affective disorders, schizophrenia (Brodie et al. 1983; Buchsbaum et al. 1982)]. Investigations as to possible abnormalities of metabolic maps of the brain in anorexia nervosa are challenging for the following three reasons:
1. In anorexia nervosa intraindividual comparison between the anorectic and remitted state of the patient can be performed.
2. Since anorectic patients reveal a very pronounced reversible pseudoatrophy of the brain during the anorectic phase, the question arises as to a metabolic functional correlate of this atrophic process in regional glucose metabolism.
3. A further topic is a possible labeling of hypothetically hypofunctional brain areas primarily responsible for the illness (e.g., hypothalamic regions relevant in hunger and satiety regulation) or other areas responsible for the complex psychobiology of this illness.

Several investigations in recent years, using conventional X-ray computer tomography, showed unequivocal evidence for a very pronounced reversible (Kohlmeyer et al. 1983) − pseudoatrophy of the brain in anorexia nervosa (Enzmann and Lane 1977; Heinz et al. 1977; Nussbaum et al. 1980; Sein et al. 1981). The very pronounced symmetrical enlargements of cortical sulci are characteristic for this cerebral atrophy, which can be interpreted as an enlargement of the subarachnoid space (prominence of the insular cisterns and of the anterior interhemispheric fissure). The pathophysiological basis underlying these abnormalities has not yet been elucidated. There are, however, some parallels with changes characteristic for Cushing's syndrome and severe malnutrition. Heinz et al. (1977) propose that protein loss and/or fluid retention are pathophysiologically relevant in the pathogenesis of these abnormalities.

---

[1] Max-Planck-Institut für Neurologische Forschung, Ostmerheimerstrasse 200, D-5000 Köln, and Max-Planck-Institut für Psychiatrie, Kraepelinstrasse 10, D-8000 München 40

From the X-ray findings summarized above, one may anticipate a hypometabolic situation in especially cortical regions of the brain during anorexia nervosa. This perspective is, furthermore, supported by the findings of Owen et al. (1967) that in anorexia nervosa — if ketonemia is present — ketone utilization also makes a sizable contribution to the brain metabolism, thereby reducing the contribution made by glucose. (To minimize this interference, only patients who exhibited near-normal levels of ketone compounds were included in the present investigation.)

On the other hand, there are some reasons to conceptualize at least normal or even increased metabolic rates of glucose metabolism in the CNS of anorectic patients, based on a functional/neuropsychological point of view: anorectic patients often appear highly alert and vivid, and have a pronounced motor hyperactivity. Though some neuropsychobiological deficits in these patients have been published (Fox 1981; Lehmkuhl et al. 1982), or at least some differences have been observed in performance comparing the anorectic and the remitted state, recent observations (D. von Cramon, 1983, unpublished data) reveal increased values in tests measuring vigilance and performance for patients with anorexia nervosa in comparison with normal controls.

The present pilot study, using $(18\text{-}F)$-2-fluoro-2-deoxy glucose PET (Reivich et al. 1979), was performed to elucidate the challenging question as to whether hyper- or hypometabolic cortical activity is present in anorexia nervosa. The study has so far been confined to investigations of the patients in the anorectic state, so that no intraindividual comparison can be performed. Hence one is restricted to comparison with a control group of normal subjects.

Methods

For PET imaging the four-ring/seven-slice system designed by Eriksson et al. (1982) was used. Cross-sectional positron images were obtained by applying the $(18\text{-}F)$-2-fluoro-2-deoxy glucose (FDG) technique described by Reivich et al. (1979). The procedures of investigation, rate constants, and lumped constant were as described by Phelps (1981). About 4 mCi FDG was injected IV, and arterialized blood was sampled from the vein of a heated hand from the time of injection until the end of the examination. From the local tracer activity at 30–40 and 40–50 min after injection and the plasma concentration of FDG and glucose, a total of 14 tomographic images of the local metabolic rates for glucose (LCMRGl) were computed, usually covering the entire brain.

Patient Data

Six patients with a moderate to severe anorexia nervosa (ICD-9 no. 307.1) were examined during the anorectic phase. They fulfilled the diagnostic criteria of Feighner et al. (1972) and of DSM-III (Spitzer 1980). Each patient gave written informed consent. The patients were drug-free, treated using a behavioral therapeutic protocol, and were inpatients in an open ward at the Max Planck Institute for Psychiatry, Munich. The PET recordings were performed at the Max Planck Institute for Neurological Research, Cologne.

The PET investigations were not started prior to near-normalization of the plasma values of free fatty acids, 3-hydroxybutyrate, and acetoacetate, to avoid a major compensatory reduction of glucose metabolism by ketone utilization.

Description of Cases

Case No. 1. A 20-year-old girl, 28% below ideal body weight (IBW) at admission (15% below IBW during PET scan), exhibiting secondary amenorrhea, bulimia, and vomiting. EEG and X-ray CT normal. Duration of illness: 2 years.

Case No. 2. A 17-year-old girl, 35% below IBW at admission (30% below IBW during PET scan), exhibiting secondary amenorrhea. No bulimia or vomiting. EEG: increased theta-activity over both hemispheres. CT: lateral ventricles slightly enlarged; cortical and cerebellar sulci enlarged. Duration of illness: 10 months.

Case No. 3. A 21-year-old girl, 40% below IBW at admission (35% below IBW during PET scan), exhibiting secondary amenorrhea, bulimia, and vomiting. EEG: normal. CT: cortical sulci enlarged. Duration of illness: 2 years.

Case No. 4. An 18-year-old girl, 35% below IBW at admission (35% below IBW during PET scan), exhibiting secondary amenorrhea, bulimia, and vomiting. EEG: slight increase of theta-activity over both hemispheres. CT: slight enlargement of lateral ventricles; cortical and cerebellar sulci enlarged. Duration of illness: 1.5 years.

Case No. 5. A 23-year-old girl, 35% below IBW at admission (30% below IBW during PET scan), exhibiting secondary amenorrhea, bulimia, and vomiting. EEG: strong increase of theta-activity over both hemispheres. CT: slight enlargement of lateral ventricles; cortical and cerebellar sulci enlarged. Duration of illness: 8 years.

Case No. 6. A 19-year-old girl, 35% below IBW at admission (35% below IBW during PET scan), exhibiting secondary amenorrhea, bulimia, and vomiting. EEG: normal. CT: cortical sulci enlarged. Duration of illness: 3 years.

Results

Figure 1 shows a comparative representation of conventional CT findings and PET metabolic maps of a patient with pronounced cortical atrophy (Fig. 1a), abnormal EEG (slight increase of theta-activity over both hemispheres), and a weight deficit of 35% below IBW (case no. 4). The metabolic maps in Fig. 1b, and, similarly those of the other five patients, appear relatively "normal". There is a high cortical activity and the fronto-occipital ratio appears not to be reduced. Figure 2 represents the preliminary data of regional metabolic rates (cross-hatched areas), showing a selection of metabolic rates of 20 areas of interest (preliminary evaluations). Intraindividual comparison has not so far been possible. Therefore, for comparison, the data from 7 normal controls are represented in Fig. 2, although these controls are not age- and sex-matched (males, on average 10 years older than the patients). Since, however, in other PET investigations no pronounced age dependence of metabolic rates has been observed, for a first evaluation this comparison may appear legitimate. Neither the patients' data nor those of the normal control subjects reveal any right/left difference. The cortical values are about the same in both groups. A tendency

PET Investigation in Anorexia Nervosa

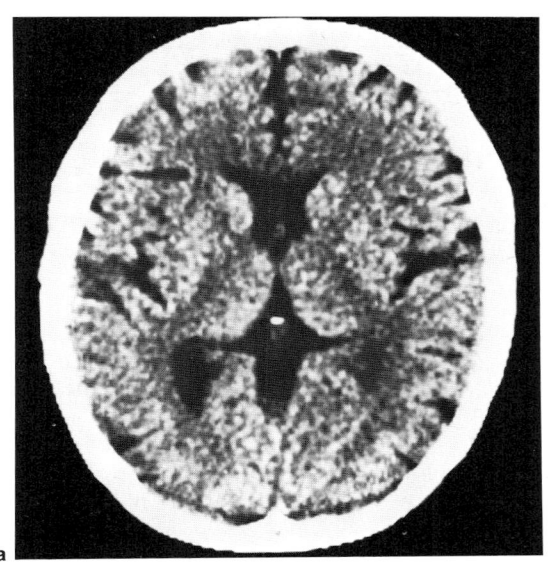

Fig. 1. a Transmission CT of patient no. 4, exhibiting pseudoatrophy of the brain. **b** Metabolic maps (FDG) of the same patient

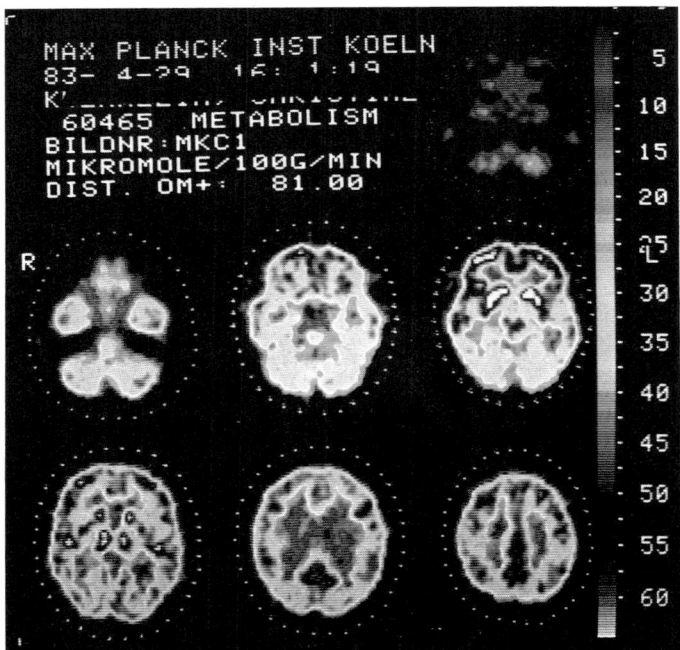

towards an abnormality may be seen in the relatively high values in the cerebellar cortex. The tendency to lower values in the insular cortex and caudate nucleus may also be of interest (cf. Table 1).

Table 1. Clinical data and metabolic rates for regions of interest in the six anorectic patients and mean

Case no.	Initials	Age	Ab-normal EEG	Ab-normal CT	Weight deficit	Metabolic rates [μmol/100 g per min]						
						Cerebellar cortex		Hippocampus		Fronto-central cortex	Fronto-medial cortex	
						R	L	R	L		R	L
1	G. Z.	20	–	–	15%	38	36	29	24	41	40	48
2	N. G.	17	+	++	30%	34	32	30	28	43	41	44
3	C. M.	21	–	+	35%	35	37	26	25	46	48	44
4	C. K.	18	(+)	++	35%	34	37	24	28	44	48	50
5	A. W.	23	++	++	30%	21	22	18	19	34	33	36
6	B. R.	19	–	+	35%	33	35	29	27	48	41	35
					x̄	33	33	26	25	43	42	43
					±	6	6	5	3	5	6	6
Normal controls					x̄	29	28	30	30	39	39	40
					±	4	4	5	6	5	5	5

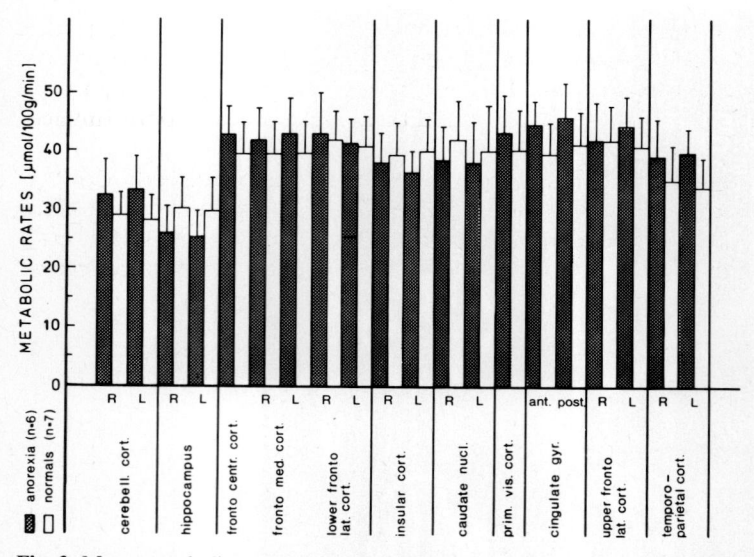

Fig. 2. Mean metabolic rates of regions of interest in the six anoretic patients compared with seven normal control subjects

Discussion

In the present investigation, FDG metabolic rates in anorectic patients during the anorectic phase have been evaluated for the first time. In contrast to the pronounced cortical pseudoatrophy, the cortical glucose metabolism of

values of seven normal controls

Lower frontolateral cortex		Insular cortex		Caudate nucleus		Primary visual cortex	Cingulate gyrus		Upper frontolateral cortex		Temporo-parietal cortex	
R	L	R	L	R	L		Ant.	Post.	R	L	R	L
37	42	34	39	33	34	47	44	51	44	45	41	42
46	47	34	35	36	36	44	44	39	39	44	34	39
46	40	40	37	42	43	41	44	44	44	46	38	42
52	41	43	33	45	43	49	47	48	44	44	44	41
32	35	32	32	31	27	31	38	39	30	35	30	32
43	45	44	41	43	45	46	50	53	49	51	47	42
43	41	38	36	38	38	43	45	46	42	44	39	40
7	4	5	4	6	7	7	4	6	6	5	6	4
42	41	39	40	42	40	40	39	41	42	41	35	34
6	5	6	6	7	8	7	5	6	6	5	6	5

anorectic patients is apparently normal. Deviations, which may exist in the cerebellar cortex (tendency towards a hypermetabolism), have to be corroborated by intraindividual comparison. The high frontal activity of the patients is in line with their high psychomotor activity and their high values in performance tests.

The question as to a possible labeling of the primary processes leading to the anorectic behavior of these patients by use of the FDG/PET technique still has to be investigated by further detailed evaluations of the data acquired so far and, especially, by intraindividual comparison of the present data with measurements during remission.

Acknowledgements. We thank the ward nurses for their invaluable help.
One of us (HME) was supported by a grant from the Deutsche Forschungsgemeinschaft (German Research Foundation).

References

Barrio JR (1983) Biochemical parameters in radiopharmaceutical design. In: Heiss W-D, Phelps ME (eds) Positron emission tomography of the brain. Springer, Berlin Heidelberg New York, pp 65–76

Brodie JD, Wolf AP, Volkow N, Christman DR, DeFina P, DeLeon M, Farkas T, Ferris SH, Fowler JS, Gomez-Mont F, Jaeger J, Russell JAG, Stamm R, Yonekura Y (1983) Evaluation of regional glucose metabolism with positron emission tomography in normal and psychiatric populations. In: Heiss W-D, Phelps ME (eds) Positron emission tomography of the brain. Springer, Berlin Heidelberg New York, pp 201–206

Buchsbaum MS, Ingvar DH, Kessler R, Waters RN, Cappelletti J, van Kammen DP, King AC, Johnson JL, Manning RG, Flynn RW, Mann LS, Bunney Jr WE, Sokoloff L (1982) Cerebral glucography with positron tomography. Use in normal subjects and in patients with schizophrenia. Arch Gen Psychiatry 39: 251–259

Enzmann DR, Lane B (1977) Cranial computed tomography findings in anorexia nervosa. J Comput Assist Tomogr 1: 410–414

Eriksson L, Bohm C, Kesselberg M, Blomqvist G, Litton J, Widen L, Bergström M, Ericson K, Greitz T (1982) A four ring positron camera system for emission tomography of the brain. IEEE Trans Nucl Sci 29: 539–543

Feighner JP, Robins E, Guze SB, Woodruff Jr RA, Winokur G, Munoz R (1972) Diagnostic criteria for use in psychiatric research. Arch Gen Psychiatry 26: 57–63

Fox CF (1981) Neuropsychological correlates of anorexia nervosa. Int J Psychiatry Med 11: 285–290

Heinz ER, Martinez J, Haenggeli A (1977) Reversibility of cerebral atrophy in anorexia nervosa and Cushing's syndrome. J Comput Assist Tomogr 1: 415–418

Kohlmeyer K, Lehmkuhl G, Poustka F (1983) Computed tomography in patients with anorexia nervosa. A J N 4: 437–438

Lehmkuhl G, Poustka F, Kohlmeyer K (1982) Clinical association to cerebral pseudoatrophy in anorexia nervosa. Paper presented at the 10th International Congress, Dublin, Ireland

Nussbaum M, Shenker IR, Marc J, Klein M (1980) Cerebral atrophy in anorexia nervosa. J Pediatr 96: 867–869

Owen OE, Morgan AP, Kemp HG, Sullivan JM, Herrera MG, Cahill GF (1967) Brain metabolism during fasting. J Clin Invest 46: 1589–1595

Phelps ME (1981) Positron computed tomography studies of cerebral glucose metabolism in man: theory and application in nuclear medicine. Semin Nucl Med 11: 32–49

Reivich M, Kuhl D, Wolf A, Greenberg J, Phelps ME, Ido T, Casella V, Fowler J, Hoffman E, Alavi A, Som P, Sokoloff L (1979) The (^{18}F) fluorodeoxyglucose method for the measurement of local cerebral glucose utilization in man. Circ Res 44: 127–137

Sein P, Searson S, Nicol AR, Hall K (1981) Anorexia nervosa and pseudo-atrophy of the brain. Br J Psychiatry 139: 257–258

Spitzer RL (ed) (1980) Diagnostic and statistical manual of mental disorders, DSM-III, 3rd edn. American Psychiatric Association, Washington DC

Concluding Remarks

K. M. Pirke and D. Ploog[1]

In the discussion following the individual contributions and in the general discussion at the end of the meeting, the questions raised in the introduction were debated in detail.

Are All Somatic Symptoms of Anorexia Nervosa a Consequence of Malnutrition?

The first problem addressed was whether all somatic symptoms of anorexia nervosa, such as disturbances in sleep and activity, endocrine malfunctions, and temperature regulation, can be considered to be merely a consequence of weight loss or whether there is evidence that some of the above-mentioned symptoms are primary signs of anorexia nervosa. It was agreed upon by all participants that for the time being there is no general answer to this question. Although there is good evidence that many symptoms of anorexia nervosa, such as sleep disturbance, increased cortisol secretion, and decreased production of triiodothyronine and gonadotropins, are merely consequences of starvation, the question as to whether other observations, such as the impaired production of vasopressin, noradrenaline and hydroxymethoxy phenylglycol (HMPG) reported at this meeting, are also caused by starvation alone remains open. Although it was stated that up to now no somatic symptoms of anorexia nervosa have clearly been shown to be a primary feature of the disease, future work should be carried out in this field. The discussion concentrated then on the problem of how this could best be achieved. Further work in dieting or starving, otherwise healthy, subjects as reported on here will yield much information with regard to the effects of acute starvation. These studies, however, are of limited value when the effects of chronic starvation, often maintained for years, are concerned. The comparison of data obtained in the anorexia nervosa patient at low body weight and after normalization of weight will be of great importance. Two precautions should be taken in interpreting such data. First, it is well known that some endocrine disturbances, such as amenorrhea, may persist for long period of time after weight gain. Second, reaching the ideal body weight (judged, for instance, according to the tables of the Metropolitan Life Insurance Company) may often not be considered to be an adequate restoration of body

---

1 Max-Planck-Institut für Psychiatrie, Kraepelinstrasse 10, D-8000 München 40

weight, especially when the patient's weight was greater than that before the onset of anorexia nervosa.

It was proposed that normalization of body weight should be assumed when the average weight of an age-matched control group is reached. Even then, it may take several years before the patient adapts to this newly found weight set point.

How Do Anorexia Nervosa and Malnutrition Influence Cerebral Metabolisms in General, and Neurotransmitters in Particular?

Preliminary studies on the glucose metabolism using positron emission tomography (PET) scan revealed no major impairment. Since this technique cannot yet be applied to the study of neurotransmitter and neuromodulator activities, we have to rely on other techniques to gain insight. One method is the measurement of specific metabolites of central neurotransmitters in the cerebrospinal fluid and other body fluids. An example of this approach is the measurement of HMPG reported here. Further insight can be expected to be gained from experiments in starving animals in which central transmitter turnover can be directly measured, as reported here.

The question as to whether our knowledge of neuroendocrine and other vegetative regulations and their disturbance in anorexia nervosa can provide us with clear information on central transmitter disturbances was discussed. It was generally agreed that for the time being, our knowledge in this field is too sparse to allow the development of a convincing hypothesis on central neurotransmitter disturbances in anorexia nervosa.

An alternative manner of research was proposed by Dr. R. Wurtman. On the basis of the well-established fact that many central neurotransmitter systems depend on the supply of essential amino acids from the plasma (see R. Wurtman and T. Wurtman, this volume), amino acid patterns after the intake of either carbohydrate or protein meals should be studied in patients with anorexia nervosa. He further suggested detailed evaluation of the amounts of carbohydrates, proteins, and fat that patients with anorexia really consume.

Is There a Vicious Circle in Anorexia Nervosa?

It has been postulated by some authors that once the patient with developing anorexia nervosa gets into a state of malnutrition, the physiological and mental consequences of this state are of great importance for sustaining the course of the disease. This vicious circle not only presents a major obstacle to gaining weight again when patients are treated; it may also be that central neurotransmitter disturbances, e.g., a decreased activity of the noradrenergic system (see Ebert et al. and Pirke et al., this volume), which persist for quite a while after weight gain, may destabilize the patient and may frequently precipitate a relapse. This view was shared by most participants.

Dr. Crisp did not agree with the concept of a vicious circle. Although he underlined the importance of consequences of starvation, such as restlessness, sleep disturbance and so on, his concept sees problem avoidance behavior as the driving force of the disease (see Crisp, this volume).

The question of pharmacological treatment of anorexia nervosa is directly related to the concept of interfering with a central nervous transmitter disturbance. Since, however, up to now our understanding of such hypothetical disturbances is very poor, the most reliable means of treatment to gain weight remains behavior therapy of one kind or another. It was generally agreed that pharmacological treatment studies have been too few and too poorly designed in the past.

Anorexia Nervosa and Depression

The last topic of discussion referred to the similarity of depression and anorexia nervosa, which has been emphasized by some recent publications concerning the frequency of depressive illness among the relatives of patients with anorexia nervosa and the depressed mood and suicidal tendencies in anorectic patients. Although the two diseases share some common symptoms, such as sleep disturbances and increased activity of the adrenal gland, the patterns of pathophysiological symptoms, especially endocrine abnormalities, are quite different.

It must be taken into account that weight loss may also occur in some patients with depressive illness. This weight loss in depressed patients may well contribute to the development of increased cortisol secretion and sleep disturbance.

The view was expressed that the "depression" in anorexia nervosa may be quite unlike the depressed mood in depressive illness. Rather, it may be a feeling of discomfort which also occurs in starving healthy subjects.

The Origins of Depression: Current Concepts and Approaches

Editor: **J. Angst**

Report of the Dahlem Workshop on The Origins of Depression: Current Concepts and Approaches, Berlin 1982, Oct. 31–Nov. 5

Rapporteurs: S. A. Checkley, H. Katschnig, W. Z. Potter, M. L. Reite, A. J. Rush

1983. 4 photographs, 12 figures, 27 tables. X, 472 pages
(Dahlem Workshop Reports, Volume 26)
ISBN 3-540-12451-9

Depression is a frequent disorder occuring in all cultures, causing much suffering and many social problems. A large number of social, psychological, and biological factors have been proposed as being causal.

This volume gives a comprehensive account of pathogenetic theories and risk factors involved in the origin of depression. In November 1982, a Dahlem Workshop devoted to interdisciplinary discussions concentrated on the following areas of research: clinical models, animal models for depression, models based on neurotransmission, neuroendocrinology, and psychophysiology as well as some other modern techniques such as PET.

This report should appeal to both researchers and therapists of depression, especially psychiatrists, psychologists, sociologists, animal researchers, psychopharmacologists, neurobiologists, and neuroendocrinologists.

F. Z. Meerson

Adaptation, Stress, and Prophylaxis

Translated from the Russian by J. Shapiro
1984. 87 figures, 61 tables. X, 329 pages
ISBN 3-540-12363-6

The mechanism of phenotypic adaptation is at present a key issue in biology and medicine. Only a profound study of this mechanism can establish the grounds for controlling the adaptation of the human organism to factors in the environment. The concept of phenotypic adaptation based on the investigations presented in this book can be applied, as will be shown, to the prevention of disease and to increasing the organism's resistance to stress-induced damage, physical loads, and other environmental factors. The book is thus aimed at a broad range of workers in medicine, biology, and related fields. At the same time, this concept naturally reflects only a certain stage in the study of this complex, and seemingly infinite problem.

The evidence prodived in this work is based on comprehensive physiologic, biochemical, and cytological studies performed at the Laboratory for Heart Pathophysiology, Institute for General Pathology and Pathological Physiology, USSR Academy of Medical Sciences, and by associated scientific groups.

Springer-Verlag
Berlin
Heidelberg
New York
Tokyo

The AMDP-System
Manual for the Assessment and Documentation of Psychopathology
Edited and translated from the German by W. Guy, T. A. Ban
In collaboration with D. Bobon et al.
1982. XII, 121 pages. ISBN 3-540-11252-9

H. B. M. Murphy
Comparative Psychiatry
The International Intercultural Distribution of Mental Illness
1982. 28 figures. IX, 327 pages. (Psychiatry Series, Volume 28) ISBN 3-540-11057-7

C. Ernst, J. Angst
Birth Order
Its Influence on Personality
With a Foreword by M. Bleuler
1983. 4 figures, 86 tables. XVII, 343 pages. ISBN 3-540-11248-0

M. Gossop
Theories of Neurosis
Foreword: H. J. Eysenck
1981. 6 figures, 4 tables. XI, 161 pages. ISBN 3-540-10370-8

H. N. Levinson
A Solution to the Riddle Dyslexia
1980. 95 figures, 9 tables. XII, 398 pages. ISBN 3-540-90515-4

Psychopathological and Neurological Dysfunctions Following Open-Heart Surgery
Editors: R. Becker, J. Katz, M.-J. Polonius, H. Speidel.
1982. 71 figures. XXIII, 384 pages. ISBN 3-540-11621-4

Psychosocial Interventions in Schizophrenia
An International View
Editors: H. Stierlin, C. Wynne, M. Wirsching
1983. 26 figures, 12 in color. XII, 251 pages. ISBN 3-540-12195-1

R. M. Torack
The Pathologic Physiology of Dementia
With Indications for Diagnosis and Treatment
1978. 11 figures, 24 tables. VIII, 155 pages (Psychiatry Series, Volume 20) ISBN 3-540-08904-7

Springer-Verlag
Berlin
Heidelberg
New York
Tokyo

MAY

DE PAUL UNIVERSITY LIBRARY
30511000142363
616.85P974P C001
LPX THE PSYCHOBIOLOGY OF ANOREXIA NERVO